Applied Catalysis in Chemical Industry: Synthesis, Catalyst Design, and Evaluation

Applied Catalysis in Chemical Industry: Synthesis, Catalyst Design, and Evaluation

Editors

Magdalena Zybert
Katarzyna Antoniak-Jurak

MDPI • Basel • Beijing • Wuhan • Barcelona • Belgrade • Manchester • Tokyo • Cluj • Tianjin

Editors

Magdalena Zybert
Faculty of Chemistry
Warsaw University of
Technology
Warsaw
Poland

Katarzyna Antoniak-Jurak
Catalyst Research Department
Łukasiewicz Research Network—
New Chemical Syntheses Institute
Puławy
Poland

Editorial Office
MDPI
St. Alban-Anlage 66
4052 Basel, Switzerland

This is a reprint of articles from the Special Issue published online in the open access journal *Catalysts* (ISSN 2073-4344) (available at: https://www.mdpi.com/journal/catalysts/special_issues/applied_catalysis_in_chemical_industry).

For citation purposes, cite each article independently as indicated on the article page online and as indicated below:

LastName, A.A.; LastName, B.B.; LastName, C.C. Article Title. *Journal Name* **Year**, *Volume Number*, Page Range.

ISBN 978-3-0365-7777-7 (Hbk)
ISBN 978-3-0365-7776-0 (PDF)

© 2023 by the authors. Articles in this book are Open Access and distributed under the Creative Commons Attribution (CC BY) license, which allows users to download, copy and build upon published articles, as long as the author and publisher are properly credited, which ensures maximum dissemination and a wider impact of our publications.

The book as a whole is distributed by MDPI under the terms and conditions of the Creative Commons license CC BY-NC-ND.

Contents

About the Editors vii

Magdalena Zybert
Applied Catalysis in Chemical Industry: Synthesis, Catalyst Design, and Evaluation
Reprinted from: *Catalysts* **2023**, *13*, 607, doi:10.3390/catal13030607 1

Wojciech Patkowski, Magdalena Zybert, Hubert Ronduda, Gabriela Gawrońska, Aleksander Albrecht, Dariusz Moszyński, et al.
The Influence of Active Phase Content on Properties and Activity of Nd_2O_3-Supported Cobalt Catalysts for Ammonia Synthesis
Reprinted from: *Catalysts* **2023**, *13*, 405, doi:10.3390/catal13020405 5

Magdalena Zybert, Aleksandra Tarka, Wojciech Patkowski, Hubert Ronduda, Bogusław Mierzwa, Leszek Kępiński and Wioletta Raróg-Pilecka
Structure Sensitivity of Ammonia Synthesis on Cobalt: Effect of the Cobalt Particle Size on the Activity of Promoted Cobalt Catalysts Supported on Carbon
Reprinted from: *Catalysts* **2022**, *12*, 1285, doi:10.3390/catal12101285 21

Aleksandra Tarka, Magdalena Zybert, Hubert Ronduda, Wojciech Patkowski, Bogusław Mierzwa, Leszek Kępiński and Wioletta Raróg-Pilecka
On Optimal Barium Promoter Content in a Cobalt Catalyst for Ammonia Synthesis
Reprinted from: *Catalysts* **2022**, *12*, 199, doi:10.3390/catal12020199 35

Paweł Adamski, Wojciech Czerwonko and Dariusz Moszyński
Thermal Stability of Potassium-Promoted Cobalt Molybdenum Nitride Catalysts for Ammonia Synthesis
Reprinted from: *Catalysts* **2022**, *12*, 100, doi:10.3390/catal12010100 49

Magdalena Saramok, Marek Inger, Katarzyna Antoniak-Jurak, Agnieszka Szymaszek-Wawryca, Bogdan Samojeden and Monika Motak
Physicochemical Features and NH_3-SCR Catalytic Performance of Natural Zeolite Modified with Iron—The Effect of Fe Loading
Reprinted from: *Catalysts* **2022**, *12*, 731, doi:10.3390/catal12070731 59

Bogdan Ulejczyk, Paweł Jóźwik, Michał Młotek and Krzysztof Krawczyk
A Promising Cobalt Catalyst for Hydrogen Production
Reprinted from: *Catalysts* **2022**, *12*, 278, doi:10.3390/catal12030278 75

Magdalena Greluk, Marek Rotko, Grzegorz Słowik, Sylwia Turczyniak-Surdacka, Gabriela Grzybek, Kinga Góra-Marek and Andrzej Kotarba
Effect of Potassium Promoter on the Performance of Nickel-Based Catalysts Supported on MnO_x in Steam Reforming of Ethanol
Reprinted from: *Catalysts* **2022**, *12*, 600, doi:10.3390/catal12060600 87

Juan C. García-Prieto, Luis A. González-Burciaga, José B. Proal-Nájera and Manuel García-Roig
Kinetic Study and Modeling of the Degradation of Aqueous Ammonium/Ammonia Solutions by Heterogeneous Photocatalysis with TiO_2 in a UV-C Pilot Photoreactor
Reprinted from: *Catalysts* **2022**, *12*, 352, doi:10.3390/catal12030352 109

Sudipto Pal, Antonietta Taurino, Massimo Catalano and Antonio Licciulli
Block Copolymer and Cellulose Templated Mesoporous TiO_2-SiO_2 Nanocomposite as Superior Photocatalyst
Reprinted from: *Catalysts* **2022**, *12*, 770, doi:10.3390/catal12070770 133

Joanna Woroszył-Wojno, Michał Młotek, Bogdan Ulejczyk and Krzysztof Krawczyk
Toluene Decomposition in Plasma–Catalytic Systems with Nickel Catalysts on CaO-Al$_2$O$_3$ Carrier
Reprinted from: *Catalysts* **2022**, *12*, 635, doi:10.3390/catal12060635 . . . . **147**

About the Editors

Magdalena Zybert

Dr. Magdalena Zybert graduated from the Faculty of Chemistry, Department of Chemical Technology of the Warsaw University of Technology, Poland, in 2010 and earned her PhD in 2015 from the same university. She is currently a member of the Technical Catalysis Group at the Warsaw University of Technology, Poland, where she conducts research on various topics related to heterogeneous catalysis, synthesis, and characterization studies of catalytic systems for industrial chemical processes, synthesis of nanomaterials, and surface science. Her main area of specialization is the development of novel catalysts for the ammonia synthesis process. Her recent research activity is in designing new cathode materials for Li-ion and Na-ion batteries. Her publications include more than 40 peer-reviewed papers in international journals and 6 patents. She is involved in numerous research projects financed by scientific institutions and the industry.

Katarzyna Antoniak-Jurak

Dr. Katarzyna Antoniak-Jurak is Head of the Catalytic Preparation Section in Łukasiewicz-INS. Her specialty is synthesizing and designing catalytic material production for hydrogen processes. She is the author of over 30 papers and 10 patent applications, and is the manager of research and R&D projects. She participates in R&D teams and cooperates with universities and business representatives in the scope of applied catalysis. She also participates in the implementation of technologies, and is a member of the Polish Chemical Society and Polish Catalysis Club.

Editorial

Applied Catalysis in Chemical Industry: Synthesis, Catalyst Design, and Evaluation

Magdalena Zybert

Faculty of Chemistry, Warsaw University of Technology, Noakowskiego 3, 00-664 Warsaw, Poland; magdalena.zybert@pw.edu.pl

Citation: Zybert, M. Applied Catalysis in Chemical Industry: Synthesis, Catalyst Design, and Evaluation. *Catalysts* **2023**, *13*, 607. https://doi.org/10.3390/catal13030607

Received: 10 March 2023
Accepted: 15 March 2023
Published: 17 March 2023

Copyright: © 2023 by the author. Licensee MDPI, Basel, Switzerland. This article is an open access article distributed under the terms and conditions of the Creative Commons Attribution (CC BY) license (https://creativecommons.org/licenses/by/4.0/).

Catalysis is a very important process with practical significance for sustainable development, energy production, environmental protection, food production, and water purification, among others, and catalytic processes produce almost 90% of the products in the chemical industry. Catalysis has great economic and strategic importance, making it a rapidly evolving field. In the face of significant contemporary challenges, it is essential to acquire fundamental knowledge about the structure and phenomena of catalytic surfaces and the relationships between a catalyst's composition, synthesis method, properties, and performance in industrial processes. Research is still needed to improve existing catalysts or to design new systems that can efficiently and selectively achieve a desired product through a given reaction. In this respect, surface science is necessary to gain deeper insight into the surface structure of catalysts, the chemical state of active sites and adsorbed species, the role of surface defects, and the mode of action of selected promoters. The combination of advanced ex situ experiments, the in situ characterisation of working catalysts, and theoretical studies leads to a better prediction of catalyst behaviour under specific process conditions and is required for the conscious design of catalysts with desirable properties.

This Special Issue contains 10 research articles devoted to designing and characterising heterogeneous catalytic systems for industrial chemical processes of high importance. The aim of this Special Issue is to collect the current state of knowledge, indicate areas that require further research, and demonstrate the direction of ongoing development work. Attention is focused on comprehensive experimental studies on the synthesis, characterisation, and performance evaluation of catalysts in various industrial processes. The Special Issue covers recent advances in the following topics: ammonia synthesis, the selective catalytic reduction of nitrogen oxides, hydrogen production from ethanol, and photocatalysis. The scope also includes an investigation of catalysts under conditions similar to industrial conditions, a comparison of the studied catalytic systems and the currently operating commercial systems, and a demonstration of the validity of their application in a given chemical process.

Hydrogen technologies are currently attracting significant attention from researchers and industry as new and prospective forms of processing and storing energy. The development of new and safe hydrogen storage and distribution methods is a critical problem that requires a solution. The method with the greatest prospect is based on bonding hydrogen with nitrogen to form an NH_3 molecule. To correspond well with a hydrogen-based economy, it is crucial to develop catalytic systems that enable more sustainable methods of ammonia production capable of being conducted under milder conditions. Research is currently focused on developing active catalytic systems based on metals other than iron. These studies mostly focus on cobalt, and obtaining a technologically interesting cobalt catalyst with satisfactory activity, proper mechanical strength, adequate stability, and a favourable price requires optimising the catalyst's composition in terms of its active phase content, type and content of promoters; and the type and properties of the support, among other factors. In this respect, Patkowski et al. [1] investigated the effect of metallic cobalt content (ranging from 10 to 50 wt.%) on the structural parameters, morphology, crystal structure, surface state, composition and activity of neodymium-oxide-supported cobalt catalysts. The activity strongly depended on the active phase's content due to the average cobalt particle size changes. The productivity per catalyst mass increased with the increase in cobalt content, while the TOF maintained constant. The TOF was only below average for the catalyst with the lowest Co content when the average Co particle size was below 20 nm. In this case, the predominance of strong hydrogen binding sites on the surface, which led to hydrogen poisoning, was observed. This effect was also related to the structural sensitivity phenomenon of the NH_3 synthesis reaction carried out on cobalt observed by Zybert et al. [2]. These studies revealed a correlation between the reactivity of the cobalt surface during ammonia synthesis and the particle size of the active phase, which was supported by activated carbon. In turn, this strongly depends on the amount of the active phase. Increasing the amount of cobalt in the range of 4.9–67.7 wt.% resulted in a significant increase in the 3–45 nm range of the Co particle size. There is an optimal cobalt particle size (20–30 nm) that ensures the highest activity of the cobalt catalyst in the ammonia synthesis reaction. The observed size effect was most likely attributed to changes in the Co crystalline structure, i.e., the appearance of the hcp Co phase (which is more active than the fcc Co phase) for particles with a diameter of 20–30 nm. The addition of selected promoters was also required to increase the catalytic activity of this metal. Tarka et al. [3] explored the influence of barium content on the physicochemical properties and activity of a cobalt catalyst doubly promoted with cerium and barium. The barium promoter's dual modes of action (structural and modifying) were revealed; however, it strictly depended on the barium-to-cerium molar ratio. For the best performance of the CoCeBa system, the Ba/Ce molar ratio should be greater than unity. This results in a structural promotion of barium and a modifying action associated with the in situ formation of the $BaCeO_3$ phase, primarily reflected in the differentiation of weakly and strongly binding sites on the catalyst's surface and changes in the cobalt surface's activity. The influence of alkali metals on ammonia synthesis catalysts is an ongoing debate. Electron transfers from the alkali to the active centres and the change in the surface structure or the crystallite sizes are often considered. In [4], Adamski et al. addressed the thermal stability problem of K-promoted cobalt molybdenum nitride catalysts as a critical factor in the practical application of catalysts. Catalysts based on the mixture of Co_3Mo_3N and Co_2Mo_3N phases were highly active in the ammonia synthesis process. Potassium promoted the catalytic activity, and the promoter content affected the phase composition. The potassium-free catalyst remained unchanged after the thermostability test, whereas the K-promoted catalysts lost their activity due to a decrease in surface area. The maximum surface area and activity were observed for the catalysts containing 0.8–1.3 wt.% of potassium, in which the concentration of the Co_2Mo_3N phase was the greatest.

Ammonia, which is easily available in nitric acid plants, is also used as a reducing agent in the selective catalytic reduction process (NH_3-SCR), where the reduction of NO_x with NH_3 to form N_2 and H_2O occurs. Saramok et al. [5] demonstrated the catalytic potential of low-cost, easy-to-prepare, iron-modified clinoptilolite in this process, which occurs within the range of 250–450 °C under near-industrial conditions, i.e., using a real tail gas mixture, which usually enters SCR reactors in nitric acid plants. The results of this study indicated that the presence of various iron species, including natural, isolated Fe^{3+} and the introduced Fe_xO_y oligomers, contributed to efficient NO_x reduction, especially in the high-temperature range in which the NO_x conversion rate exceeded 90%. The concentration of N_2O also decreased by 20% when compared to its initial concentration.

Effective methods for producing hydrogen from renewable resources are also being intensively investigated. The steam reforming of ethanol is one such promising method. Ulejczyk et al. [6] reported the development of a cobalt catalyst that produced hydrogen from a mixture of ethanol and water. This work revealed that excess water is beneficial and increases the concentration of H_2 and CO_2. The metal catalyst was resistant to sintering and active phase migration, enabling a high ethanol conversion and a high hydrogen production efficiency. As carbon deposition is the primary deactivation mechanism for catalysts in the ethanol steam reforming process, Greluk et al. [7] investigated the role of a potassium promoter on the stability and resistance to carbon deposit formation of nickel-based catalysts supported on MnO_x. The studies revealed that by modifying Ni/MnO_x with potassium, the catalyst stability can be improved for different $H_2O/EtOH$ molar ratios under SRE conditions. The promoter inhibits the accumulation of carbon on the catalyst's surface, an effect resulting from the presence of potassium on the Ni surface. This leads to a decrease in the number of active sites available for methane decomposition and an increase in the rate of formed carbon's steam gasification.

In [8,9], the importance of designing reactors and catalysts for photocatalytic processes, which are essential for treating urban and industrial wastewater, was emphasized. García-Prieto et al. [8] conducted a kinetic study and modelled the heterogeneous photocatalytic degradation of aqueous ammonium/ammonia solutions using SiO_2-supported TiO_2 in a pilot UV-C photoreactor; they considered both photolytic and photocatalytic processes. This paper presents a sensitivity analysis of the main variables and mechanisms for ammonia removal. It was stated that NH_4^+/NH_3 can be decomposed by both photolysis and photocatalysis routes. However, the UV-C photocatalytic process was more effective for degrading NH_4^+/NH_3 than the photolytic process. Pal et al. [9] demonstrated a unique templating method to anchor the TiO_2-SiO_2-triblock copolymer composite on a cellulose matrix and obtained a pure, inorganic TiO_2-SiO_2 mesoporous nanostructure after burning out the cellulose template and the copolymer. The triblock copolymer acted as a structure-directing and mesopore-generating agent, whereas the SiO_2 counterpart fixed the thermal stability of the anatase phase and the mesostructure, and the cellulose templating enhanced the specific surface area and porosity. The TiO_2-SiO_2 nanocomposites showed excellent thermal stability in the anatase phase and a much higher photocatalytic efficiency than the TiO_2/SiO_2 without the cellulose templating, as did the standard reference catalyst, P25 TiO_2.

Lastly, Woroszył-Wojno et al. [10] studied the problem of efficiently removing tar from the gas obtained after biomass pyrolysis. They used plasma–catalysis systems with nickel catalysts deposited on Al_2O_3 and CaO-Al_2O_3. For the $NiO/(CaO$-$Al_2O_3)$ catalyst, high conversions of toluene (as a tar imitator) were observed—up to 85%—which exceeded the results obtained without the catalyst. The products of the toluene decomposition reactions were not adsorbed onto its surface. The calorific value of the outlet gas was unchanged during the process and was higher than required for turbines and engines.

In summary, the articles published in this Special Issue are an excellent representation of the advances and current development directions in industrial catalysis. They emphasize the importance of progress in designing new catalysts and reactors for the advanced chemical industry. I would like to express my gratitude to the authors for their valuable contributions and to the reviewers for their time and many constructive comments, which helped improve the quality of the published papers. Special thanks to the Editorial Office of *Catalysts* for allowing me to serve as a Guest Editor. I hope that this Special Issue will greatly interest a broad group of readers, especially those involved in topics related to the synthesis and characterisation of catalytic systems for industrial chemical processes, heterogeneous catalysis, and surface science.

Conflicts of Interest: The author declares no conflict of interest.

References

1. Patkowski, W.; Zybert, M.; Ronduda, H.; Gawrońska, G.; Albrecht, A.; Moszyński, D.; Fidler, A.; Dłużewski, P.; Raróg-Pilecka, W. The Influence of Active Phase Content on Properties and Activity of Nd_2O_3-Supported Cobalt Catalysts for Ammonia Synthesis. *Catalysts* **2023**, *13*, 405. [CrossRef]
2. Zybert, M.; Tarka, A.; Patkowski, W.; Ronduda, H.; Mierzwa, B.; Kępiński, L.; Raróg-Pilecka, W. Structure Sensitivity of Ammonia Synthesis on Cobalt: Effect of the Cobalt Particle Size on the Activity of Promoted Cobalt Catalysts Supported on Carbon. *Catalysts* **2022**, *12*, 1285. [CrossRef]
3. Tarka, A.; Zybert, M.; Ronduda, H.; Patkowski, W.; Mierzwa, B.; Kępiński, L.; Raróg-Pilecka, W. On Optimal Barium Promoter Content in a Cobalt Catalyst for Ammonia Synthesis. *Catalysts* **2022**, *12*, 199. [CrossRef]

4. Adamski, P.; Czerwonko, W.; Moszyński, D. Thermal Stability of Potassium-Promoted Cobalt Molybdenum Nitride Catalysts for Ammonia Synthesis. *Catalysts* **2022**, *12*, 100. [CrossRef]
5. Saramok, M.; Inger, M.; Antoniak-Jurak, K.; Szymaszek-Wawryca, A.; Samojeden, B.; Motak, M. Physicochemical Features and NH$_3$-SCR Catalytic Performance of Natural Zeolite Modified with Iron—The Effect of Fe Loading. *Catalysts* **2022**, *12*, 731. [CrossRef]
6. Ulejczyk, B.; Jóźwik, P.; Młotek, M.; Krawczyk, K. A Promising Cobalt Catalyst for Hydrogen Production. *Catalysts* **2022**, *12*, 278. [CrossRef]
7. Greluk, M.; Rotko, M.; Słowik, G.; Turczyniak-Surdacka, S.; Grzybek, G.; Góra-Marek, K.; Kotarba, A. Effect of Potassium Promoter on the Performance of Nickel-Based Catalysts Supported on MnO$_x$ in Steam Reforming of Ethanol. *Catalysts* **2022**, *12*, 600. [CrossRef]
8. García-Prieto, J.C.; González-Burciaga, L.A.; Proal-Nájera, J.B.; García-Roig, M. Kinetic Study and Modeling of the Degradation of Aqueous Ammonium/Ammonia Solutions by Heterogeneous Photocatalysis with TiO$_2$ in a UV-C Pilot Photoreactor. *Catalysts* **2022**, *12*, 352. [CrossRef]
9. Pal, S.; Taurino, A.; Catalano, M.; Licciulli, A. Block Copolymer and Cellulose Templated Mesoporous TiO$_2$-SiO$_2$ Nanocomposite as Superior Photocatalyst. *Catalysts* **2022**, *12*, 770. [CrossRef]
10. Woroszył-Wojno, J.; Młotek, M.; Ulejczyk, B.; Krawczyk, K. Toluene Decomposition in Plasma–Catalytic Systems with Nickel Catalysts on CaO-Al$_2$O$_3$ Carrier. *Catalysts* **2022**, *12*, 635. [CrossRef]

Disclaimer/Publisher's Note: The statements, opinions and data contained in all publications are solely those of the individual author(s) and contributor(s) and not of MDPI and/or the editor(s). MDPI and/or the editor(s) disclaim responsibility for any injury to people or property resulting from any ideas, methods, instructions or products referred to in the content.

Article

The Influence of Active Phase Content on Properties and Activity of Nd₂O₃-Supported Cobalt Catalysts for Ammonia Synthesis

Wojciech Patkowski [1], Magdalena Zybert [1,*], Hubert Ronduda [1], Gabriela Gawrońska [1], Aleksander Albrecht [2], Dariusz Moszyński [2], Aleksandra Fidler [3], Piotr Dłużewski [3] and Wioletta Raróg-Pilecka [1,*]

[1] Faculty of Chemistry, Warsaw University of Technology, Noakowskiego 3, 00-664 Warsaw, Poland
[2] Faculty of Chemical Technology and Engineering, West Pomeranian University of Technology in Szczecin, Pułaskiego 10, 70-322 Szczecin, Poland
[3] Institute of Physics, Polish Academy of Sciences, Al. Lotników 32/46, 02-668 Warsaw, Poland
* Correspondence: magdalena.zybert@pw.edu.pl (M.Z.); wioletta.pilecka@pw.edu.pl (W.R.-P.)

Citation: Patkowski, W.; Zybert, M.; Ronduda, H.; Gawrońska, G.; Albrecht, A.; Moszyński, D.; Fidler, A.; Dłużewski, P.; Raróg-Pilecka, W. The Influence of Active Phase Content on Properties and Activity of Nd₂O₃-Supported Cobalt Catalysts for Ammonia Synthesis. *Catalysts* **2023**, *13*, 405. https://doi.org/10.3390/catal13020405

Academic Editors: Jin-Hyo Boo and Fan Yang

Received: 21 December 2022
Revised: 7 February 2023
Accepted: 11 February 2023
Published: 14 February 2023

Copyright: © 2023 by the authors. Licensee MDPI, Basel, Switzerland. This article is an open access article distributed under the terms and conditions of the Creative Commons Attribution (CC BY) license (https://creativecommons.org/licenses/by/4.0/).

Abstract: A series of neodymium oxide-supported cobalt catalysts with cobalt content ranging from 10 to 50 wt.% was obtained through the recurrent deposition-precipitation method. The effect of active phase, i.e., metallic cobalt, content on structural parameters, morphology, crystal structure, surface state, composition and activity of the catalysts was determined after detailed physicochemical measurements were performed using ICP-AES, N₂ physisorption, XRPD, TEM, HRTEM, STEM-EDX, H₂-TPD and XPS methods. The results indicate that the catalyst activity strongly depends on the active phase content due to the changes in average cobalt particle size. With the increase of the cobalt content, the productivity per catalyst mass increases, while TOF maintains a constant value. The TOF is below average only for the catalyst with the lowest cobalt content, i.e., when the average Co particle size is below 20 nm. This is due to the predominance of strong hydrogen binding sites on the surface, leading to hydrogen poisoning which prevents nitrogen adsorption, thus inhibiting the rate-determining step of the process.

Keywords: cobalt catalyst; ammonia synthesis; neodymium oxide; support; hydrogen poisoning

1. Introduction

The world's growing population requires more and more food to survive. Increased consumption forces the need to intensify agricultural production based on mineral fertilisers. The primary raw material used to produce mineral fertilisers is ammonia, a nitrogen source. The latest estimates indicate that global ammonia production is increasing by more than 2% annually. In 2020 it reached 150 million tonnes, contributing to the consumption of nearly 2% of the world's energy generated from fossil fuels [1]. Ammonia production is a large-scale catalytic industrial process with significant energy consumption. The compound is obtained directly from gaseous hydrogen and nitrogen [2]. The equilibrium of the reaction requires the use of high temperature and high pressure to obtain economically viable amounts of the product. Due to the high activation energy of the nitrogen molecule, it is necessary to use a catalyst to obtain a sufficiently high reaction rate. Currently, the most used reaction catalyst is a system based on metallic iron, working effectively at a temperature of 400 °C–500 °C and under high pressure, reaching up to 30 MPa in older installations [3]. One of the critical areas of process optimisation is the alleviation of reaction conditions. For this purpose, new catalysts capable of operating effectively under lower pressure and at lower temperatures are designed [4]. Research is focused on developing active catalytic systems based on metals other than iron.

Cobalt, a metal indicated by the volcanic curve, has a high potential for catalysing ammonia synthesis reactions [5]. Starting from the studies of Hagen et al. in 2002 [6,7], a continuously increasing interest in cobalt as the active phase of ammonia synthesis catalysts has been observed. However, obtaining a technologically interesting cobalt catalyst, i.e., one with a satisfactory activity, proper mechanical strength, adequate stability and favourable price, requires optimisation of the catalyst composition in terms of (i) type and properties of the support, (ii) type and content of promoters and (iii) active phase content. Support allows for better use of the catalyst's potential by improving the dispersion of the active phase, stabilising the metal particles on the catalyst surface, influencing the morphology and size of the metal particles and reducing the cost of catalyst production by lowering the amount of the active phase used. This is crucial due to the high and fluctuating price of cobalt as a strategic metal with limited deposits located mainly in politically unstable regions. So far, literature has reported the use of activated carbons [8,9], cerium oxide [10–12] and recently also magnesium oxide [13] or mixed $MgO-Ln_2O_3$ oxides (Ln = La, Nd, Eu) [14–18] as support for cobalt catalyst. Recent reports also point to such materials as electrides (e.g., $C12A7:e^-$) [19] or hydride support materials (e.g., BaH_2) [20] as very effective support promoting ammonia synthesis over a Co-based catalyst. Because cobalt itself is not highly active in the synthesis of ammonia, the addition of selected promoters is also required to increase the catalytic activity of this metal. Barium [21–24] and rare earth metals (especially cerium [25] and lanthanum [26]) must be mentioned among the most effective promoters of cobalt.

The amount of the active phase deposited on the support is also essential. In the case of ammonia synthesis reactions carried out on metals such as cobalt, the reaction's structural sensitivity should be considered. Our previous studies revealed the correlation between the reactivity of the cobalt surface (expressed as TOF) in ammonia synthesis and the particle size of the active phase supported on activated carbon [27]. This, in turn, strongly depends on the amount of active phase. There is an optimal size of cobalt particles (20–30 nm), which ensures the highest activity of the cobalt catalyst in the ammonia synthesis reaction. Increasing or decreasing the particle size caused a decrease in activity, even to the total loss of catalyst activity for fine Co particles (smaller than 0.5 nm). The observed size effect was most likely attributed to changes in Co crystalline structure [27]. However, metal-support interactions and the effect of the amount of active phase loaded on the support on its particle size as well as exposed surface area, and catalytic activity, are usually specific to each catalytic system. In the case of the carbon support, increasing the amount of cobalt introduced to the range of 4.9–67.7 wt.% resulted in a significant increase in the size of cobalt particles in the range of 3–45 nm [27]. However, when cobalt was deposited on the mixed $MgO-La_2O_3$ support in the amount of 10–50 wt.%, the size and structure of Co nanoparticles in the catalysts remained nearly unchanged despite the fivefold increase in the Co loading amount [28].

The study presented in this paper aimed to investigate the influence of active phase, i.e. metallic cobalt content on the properties and activity of Nd_2O_3-supported cobalt catalysts for ammonia synthesis. As reported in the previous studies [29–32], rare earth metal oxides were effective supports of the ruthenium catalyst for ammonia synthesis. Niwa et al. [29] showed that the use of rare-earth metal oxides as supports is more effective than using their cations in the role of promoters. Ruthenium catalysts deposited on rare-earth metal oxides were almost twice as active as the reference systems (Ru/MgO) promoted with the rare-earth metal cations. The increased activity was attributed to Strong Metal-Support Interactions (SMSI). Miyahara et al. [30] revealed different activities of ruthenium catalysts depending on the support used, with higher activity characterising catalysts deposited on lighter oxides according to the following trend: $Pr_2O_3 > CeO_2 > La_2O_3 > Nd_2O_3 > Sm_2O_3 > Gd_2O_3$. The high activity of the Ru/Pr_2O_3 system was explained by Sato et al. [31] due to the favourable morphology of the catalyst surface in the form of a ruthenium nanolayer rich in defects and terraces, structurally similar to the active B5 sites. Additionally, the high alkalinity of the support was conducive to effective charge transfer to the metal surface and facilitated the dissociation of the adsorbed N_2 molecule (a rate-determining step of the NH_3 synthesis reaction). During kinetic studies of the Ru/Pr$_2$O$_3$ catalyst, Imamura et al. [32] also indicated high resistance to hydrogen and product poisoning, unlike Ru systems based on carbon or oxide supports. These studies were an inspiration to use rare-earth metal oxides as supports for cobalt catalysts for ammonia synthesis. Neodymium oxide was also selected based on our previous studies [18], where Nd$_2$O$_3$ had the most favourable effect on the catalytic properties of cobalt systems supported on mixed MgO-Ln$_2$O$_3$ oxides (where Ln = La, Nd, Eu).

In the present work, a series of cobalt catalysts containing 10–50 wt.% of cobalt supported on neodymium oxide was prepared. The activity of the catalysts with different Co loading was tested in ammonia synthesis at 470 °C under the pressure of 6.3 MPa. The detailed characterisation studies using N$_2$ physisorption, X-ray powder diffraction (XRPD), microscopic methods (TEM, HRTEM, STEM-EDX), X-ray photoelectron spectroscopy (XPS) and temperature-programmed desorption (H$_2$-TPD) were conducted to understand the effect of active phase content on catalyst structural parameters, morphology, crystal structure and surface state, as well as composition and activity of the catalysts.

2. Results and Discussion

Table 1 lists the textural parameters (specific surface area, S$_{BET}$, and the total volume of pores, V$_{por}$) of catalyst precursors determined using N$_2$ physisorption. All systems display a relatively developed specific surface area. Increasing the cobalt loading causes an expansion of the S$_{BET}$ area of the catalyst precursor. In the case of subsequent systems, the specific surface area increase is logarithmical, yielding diminishing growth for every additional wt.% increase in the Co content of the precursor. The porosity of the precursors follows the growth trend of the surface area.

Table 1. Textural parameters of cobalt catalyst precursors.

Parameter	Co(10)/Nd$_2$O$_3$	Co(19)/Nd$_2$O$_3$	Co(31)/Nd$_2$O$_3$	Co(39)/Nd$_2$O$_3$	Co(50)/Nd$_2$O$_3$
Specific surface area S$_{BET}$ [1] [m^2 g^{-1}]	27.3	32.0	36.5	39.6	41.8
Total pore volume V$_{por}$ [2] [cm^3 g^{-1}]	0.104	0.125	0.132	0.139	0.142

[1]—estimated based on the BET isotherm. [2]—estimated based on the BJH isotherm.

Figure 1a depicts the N$_2$ physisorption isotherms registered for catalyst precursors. All registered curves are of type II shape, indicating that precursors are predominantly macroporous materials. Increasing cobalt content does not change the isotherm shape. However, they shift upward to higher total adsorbate volumes. This observation, combined with the fact that the Nd$_2$O$_3$, used as support is non-porous and of low surface area (ca. 2 m^2 g^{-1}), indicates that the porosity may be attributed to the structures formed by cobalt oxide. Therefore, depositing subsequent layers of cobalt compounds on the surface of the output catalyst precursor leads to the expansion of the existing porous structure without changing its nature. Also, all curves contain a hysteresis loop of H3 type caused by capillary condensation in mesopores. Its presence suggests that cobalt oxide forms the aggregates of plate-like particles giving rise to wedge- or slit-shaped pores. With increasing cobalt content in the precursor, the area of the loop increases, indicating an intensification of the capillary condensation phenomenon, thus increasing mesopore total volume. Figure 1b depicts the pore volume distribution of the precursors. The curves indicate that these are porous materials with bimodal pore volume distribution. The porous structure is formed by numerous pores in the 20–80 nm diameter range and macropores with sizes above 80 nm. The materials contain a very small number of micropores.

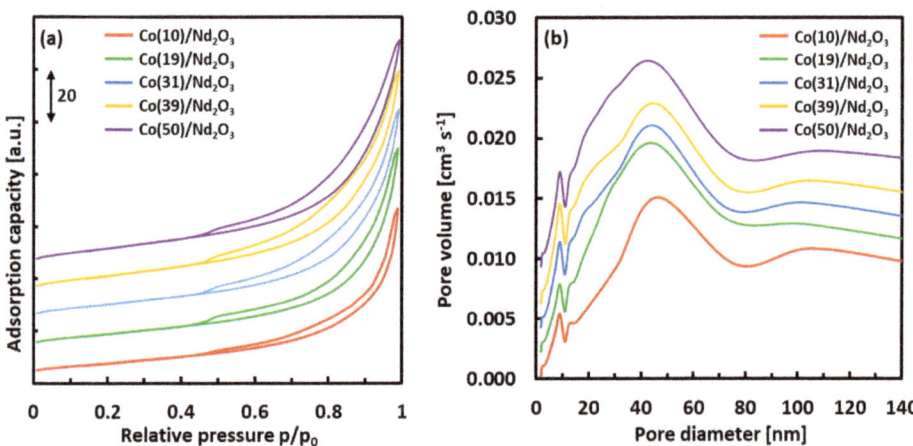

Figure 1. N$_2$ physisorption isotherms (**a**) and pore volume distribution (**b**) of cobalt catalyst precursors.

X-ray powder diffraction studies (XRPD) were carried out to determine the phase composition of catalysts. Figure 2 depicts patterns of cobalt catalyst in the form of a precursor and in-situ reduced form.

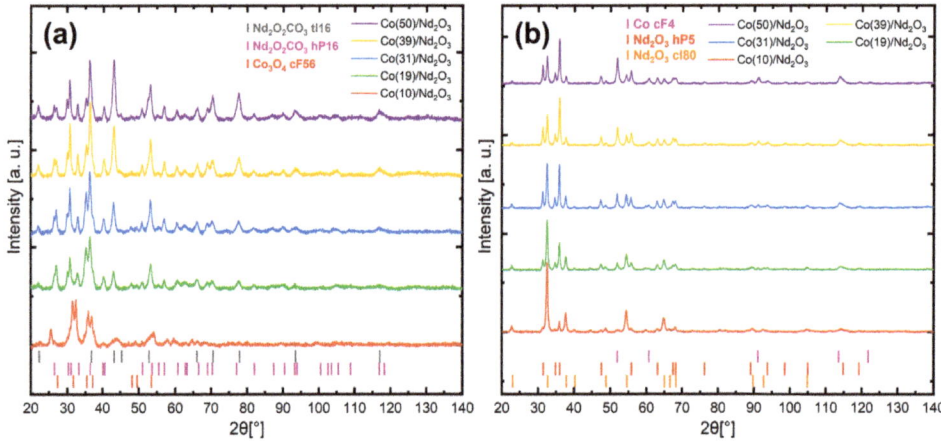

Figure 2. XRPD patterns of the cobalt catalysts supported on neodymium oxide (Co/Nd$_2$O$_3$) in the form of a precursor (**a**) and the in-situ reduced form (**b**).

The phase composition of all catalyst precursors (Figure 2a) is very similar. On most patterns, one can observe distinct Bragg's reflections indicating the presence of Co$_3$O$_4$ spinel with a cubic structure (PDF-4+ 2021 04-003-0984). The presence of this phase cannot be unambiguously demonstrated for the Co(10)/Nd$_2$O$_3$ system due to the overlapping of reflection profiles. The intensity of reflections from the Co$_3$O$_4$ phase for systems with increasing cobalt content increases. The calculated mean crystallite size of the Co$_3$O$_4$ phase is generally independent of the cobalt content and is in the range of 11–12 nm. No signals from Nd$_2$O$_3$ (the support) are observed on the precursor patterns. Instead, reflections attributed to neodymium dioxycarbonates Nd$_2$O$_2$CO$_3$ with tetragonal (PDF-4+ 2021 00-025-0567) and hexagonal structure (PDF-4+ 2021 04-009-3412) are visible. This is due to the conditions under which the catalyst precursors were synthesised. The high-temperature water environment rich in CO$_3^{2-}$ ions enabled the formation of dioxycarbonates [33,34].

Figure 2b depicts the diffraction profiles of in-situ reduced Co/Nd_2O_3 catalysts at ambient temperature. As a result of the reduction process, a disappearance of reflections attributed to the Co_3O_4 spinel occurs. This results in reflections from the cobalt metallic phase of the cubic face-centred structure (PDF-4+ 2021 01-077-7452) present in the patterns. A detailed description of these reflections is difficult because their location coincides with numerous reflections generated by the support phases. Increasing the cobalt loading in the catalyst leads to an increase in the metallic Co phase concentration and thus intensities of Co reflections. All measurements showed similar cobalt mean crystallite size resulting from analysis based on the Rietveld method. As a result of catalyst reduction, the complete disappearance of reflections attributed to polymorphic $Nd_2O_2CO_3$ is also observed. The decomposition of dioxycarbonates leads to the exposure of support structures, in this case Nd_2O_3 [35]. In Figure 2b, sets of reflections are observed, indicating the presence of two phases of neodymium oxide with different crystal structures in the material: regular (PDF-4+ 2021 03-065-3184) and hexagonal (PDF-4+ 2021 01-072-8425).

The phase analysis presented above indicates that cobalt is present as cobalt oxide Co_3O_4 in the precursors, which is then completely reduced to metallic cobalt during the precursor activation. In the case of other similar catalytic systems in which cobalt was deposited on the surface of oxide support, such as the Co/Mg-La system [14,28], incomplete reduction of cobalt compounds contained in the catalyst were observed. For this reason, the surface composition of the catalysts in both precursor and active forms was investigated by X-ray photoelectron spectroscopy (XPS).

The presence of cobalt, neodymium, oxygen and slight adventitious carbon contamination was indicated on the surface of the precursors. After the reduction process, cobalt, neodymium and oxygen remained on the surface. Figure 3 shows the XPS spectrum of the Co 2p detailed region for the precursor and active form of the $Co(10)/Nd_2O_3$ catalyst. The figure also shows the result of the deconvolution of the XPS spectra for both samples, based on the method presented in the work of Biesinger et al. [36]. The components of the XPS Co 2p lines originating from the precursor are marked with thin lines. The components coming from the catalyst after the precursor reduction process are highlighted as grey areas. The location of the maximum of the Co $2p_{3/2}$ peak at a binding energy of 780 eV indicates the presence of cobalt oxides on the surface of the precursor. In the binding energy region from 784 eV to 793 eV, characteristic satellites appear in the spectrum of the Co 2p detailed region. Based on the envelope shape of the Co 2p peaks, the presence of Co_3O_4 oxide can be ascertained [36], previously confirmed by analysis based on X-ray diffraction studies.

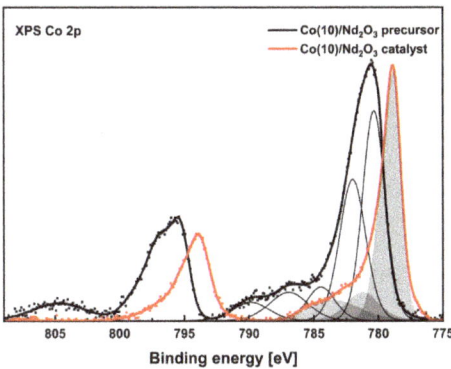

Figure 3. XPS Co 2p spectra of $Co(10)/Nd_2O_3$ precursor and catalyst after reduction. The components of the Co 2p lines are marked with thin solid lines for the precursor and grey areas for the reduced catalyst.

The exposure of precursors to hydrogen at elevated temperatures results in the process of reduction of cobalt oxides [37]. In Figure 3, a shift in the maximum of the Co 2p line from the 780 eV position to 778 eV, corresponding to the reduction of cobalt oxides to metallic cobalt, is observed. No surface cobalt oxides are observed in any samples analysed after their reduction process in hydrogen. Therefore, it should be concluded that cobalt exists exclusively in metallic form in the tested catalysts, which was confirmed by X-ray studies.

The selected catalysts of the lowest (Co(10)/Nd$_2$O$_3$), medium (Co(31)/Nd$_2$O$_3$) and highest (Co(50)/Nd$_2$O$_3$) cobalt content were examined using transmission electron microscopy (TEM), high-resolution transmission electron microscopy (HRTEM), scanning transmission electron microscopy (STEM) and energy dispersive X-rays (EDX) analysis. These examinations were used to obtain information about the morphology of the systems, the size distribution of the cobalt particles and their crystal structure.

TEM images registered for the cobalt catalysts (Figure 4) indicate that all systems display similar structures and morphology. Most materials are homogenous and form agglomerates of particles of size ranging from 20 to 100 nm. These agglomerates consist of particles that are not tightly packed together, and distances among them do not exceed 100 nm. These results agree with the pore size distribution obtained through the N$_2$ physisorption measurements, assuming the morphology of the samples does not change significantly due to the reduction process.

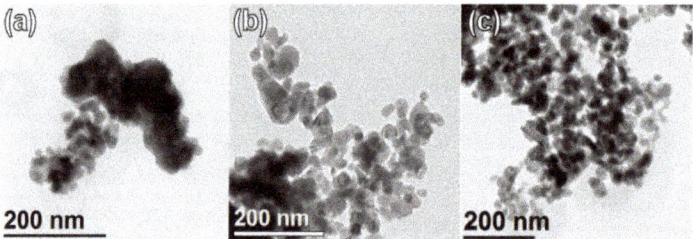

Figure 4. TEM images of regions of the Co(10)/Nd$_2$O$_3$ (**a**), Co(31)/Nd$_2$O$_3$ (**b**), Co(50)/Nd$_2$O$_3$ (**c**) catalysts.

Figure 5 depicts the HRTEM images of Co crystallites in the selected cobalt catalysts. The structure of cobalt crystallites was identified based on the analysis of the 2D-FT images from the selected areas. It was determined that for all cobalt crystals observed, the interplanar distances and the angle between them correspond to 0.207 nm and 70.5°, respectively, describing the (-111) and (-11-1) planes of the cobalt face-centred cubic structure (*Fm-3m* space group). The structure in question (ICSD 760020) is characterised by the unit cell parameter *a* = 0.3578 nm. Figure 5a also depicts that the observed Co crystallite in the Co(10)/Nd$_2$O$_3$ system is covered with a thin (ca. 2 nm) layer of CoO. The presence of the layer results from the partial oxidation of the Co crystallite caused by the nature of the ex-situ measurements. Based on the HRTEM and FT images from area no. 2, it was found that the interplanar distances and the angle between them are equal to 0.246 nm, 0.213 nm and 54.7°, respectively, corresponding to the (1-11) and (002) planes of the face-centred cubic CoO structure (space group *Fm-3m*). The unit cell parameter for the CoO structure (ICSD 245324) *a* = 0.4264 nm. The phenomenon of partial surface oxidation was also observed in the case of several Co crystallites in the other systems that underwent HRTEM investigation. It is worth noting that the face-centred cubic structure is the only type of cobalt structure observed in high-resolution images of all Co/Nd$_2$O$_3$ systems, which is consistent with the XRPD analysis results.

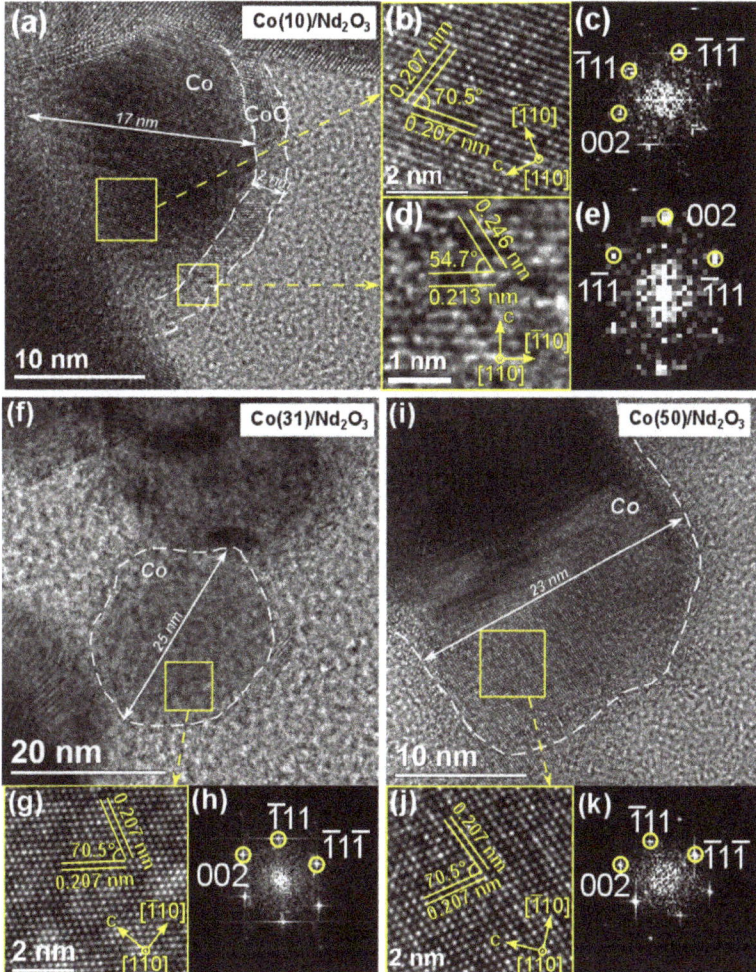

Figure 5. HRTEM images of exemplary Co crystallites (**a,f,i**) and magnified images of highlighted regions of the Co crystallites (**b,g,j**) and CoO layer (**c**) with corresponding Fourier transform (FT) (**d,e,h,k**). Yellow circles mark the reflections on the FT generated by the indicated crystal planes.

The composition and distribution of elements in the Co/Nd$_2$O$_3$ systems were obtained by STEM imaging coupled with EDX analysis. Mapping results are presented in Figure 6. The images show that for the Co(10)/Nd$_2$O$_3$ system, single Co particles are uniformly distributed over the Nd$_2$O$_3$ support, but their random agglomeration can be observed. With the increase of the Co loading in the catalyst, the particle distribution becomes less uniform. For the Co(31)/Nd$_2$O$_3$ system, Co particles are very densely distributed in some areas of the support surface, forming clusters exceeding 100 nm in size. In other regions, however, single particles of the active phase can be distinguished. For the Co(50)/Nd$_2$O$_3$ system, the cobalt particle compaction with the increasing cobalt loading continues and they are very densely distributed on the Nd$_2$O$_3$ support. Their distribution, however, is quite uniform; most of the particles form tightly packed agglomerates with only a few individual Co particles visible. With the increase in the cobalt content, the average size of metal particles increases. This observation, combined with the visible agglomeration, may indicate that Nd_2O_3 does not exert a structural influence on the cobalt phase. In contrast, the effect of prevention of particles from agglomeration and sintering was observed for the other rare-earth oxides used as supports or promoters of cobalt catalysts, namely La_2O_3 [26,38] and CeO_2 [21,25,39].

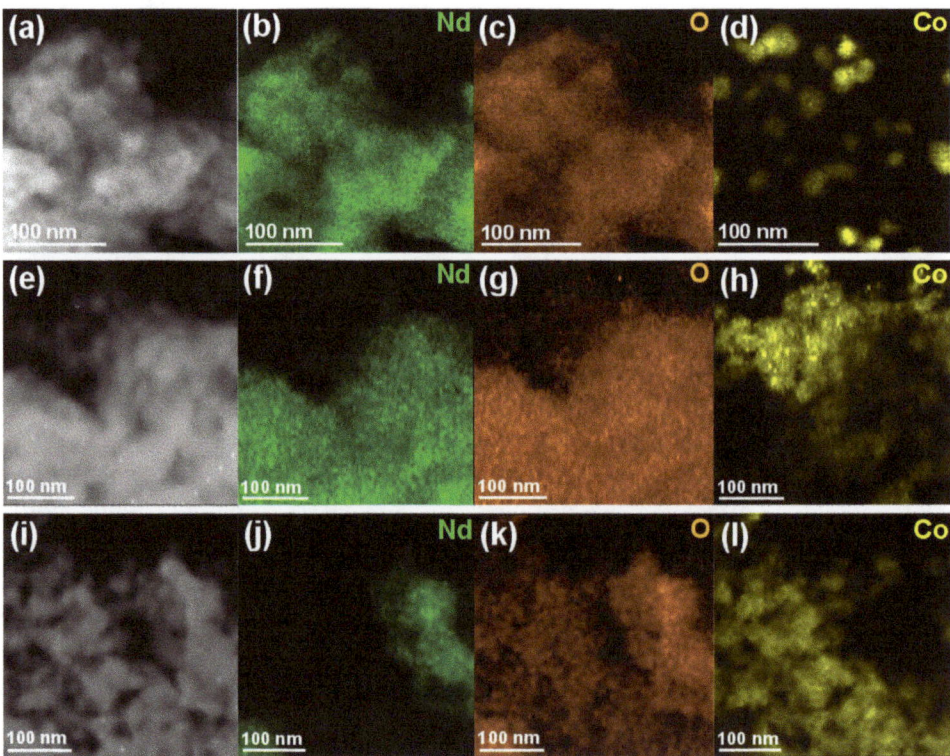

Figure 6. STEM images (first column) and corresponding elemental EDX mappings of the Co(10)/Nd_2O_3 (**a–d**), Co(31)/Nd_2O_3 (**e–h**), Co(50)/Nd_2O_3 (**i–l**) catalysts, showing Nd (green), O (orange) and Co (yellow) concentrations.

Figure 7 depicts hydrogen desorption profiles recorded for reduced Co/Nd_2O_3 catalysts. A very similar bimodal character describes all registered desorption profiles. Both low- (LT) and high-temperature (HT) peaks can be distinguished in each profile, the area of which gradually changes between systems with increasing active phase content. The first observed signal contributing to the formation of an atypical low-temperature peak is a small local signal occurring at 50 °C, corresponding to the desorption of hydrogen weakly bound to the catalyst surface. The main low-temperature peak, corresponding to the presence of adsorption centres with low hydrogen binding energy, consists of overlapping peaks with maxima located at 150 °C for the Co(10)/Nd_2O_3 system and at 100 °C for the others. The intensity of these signals increases significantly for subsequent systems in the series, along with the increasing content of the active phase. The irregular shape of the low-temperature peak indicates some variation in the hydrogen binding centres, all with relatively low binding energy [39,40]. The low-temperature signal disappears and systematically shifts into a high-temperature signal at 400 °C–500 °C. The signal above this temperature consists of a single wide peak indicating the presence of high-energy binding sites on the catalyst surface. Its area and maximum temperature differ between the systems. In the case of the Co(10)/Nd_2O_3 profile, its maximum is located at 715 °C. The peak has the largest area and a symmetrical shape. The high-temperature signal of other systems is weaker, which, when combined with the lack of such a clear maximum, may suggest fewer centres strongly binding hydrogen. A temperature shift of the maximum of the high-energy peak is also observed from 765 °C for Co(19)/Nd_2O_3 to approximately 700 °C for the Co(50)/Nd_2O_3 system, indicating the weakening of their average strength.

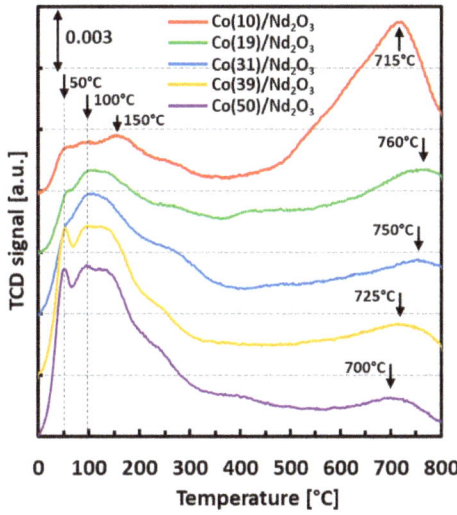

Figure 7. H_2-TPD profiles for Co/Nd_2O_3 catalysts.

Table 2 presents the measured volume of hydrogen desorbed from the catalyst surface and the share of low and high-temperature signals in the total volume. While the total volume of desorbed H_2 does not vary significantly between the samples and oscillates in the range of 1.7–2 $cm^3\ g^{-1}$, the share of the particular signals changes notably. With the increase of Co loading, the high-temperature peak share decreases and that of the low-temperature peak increases. Consequently, for the Co(10)/Nd_2O_3, the majority of the hydrogen desorbed is generated by the strong binding sites, while for the Co(50)/Nd_2O_3, the hydrogen comes mostly from weakly binding sites. Table 3 presents the average cobalt (i.e., the active phase) particle size determined through H_2-TPD and a comparison of these values to TEM and XRPD-derived data.

Table 2. The total volume of hydrogen desorbing from the catalyst surface and the proportion of low- and high-temperature peak contribution.

Parameter	Co(10)/Nd_2O_3	Co(19)/Nd_2O_3	Co(31)/Nd_2O_3	Co(39)/Nd_2O_3	Co(50)/Nd_2O_3
Total hydrogen volume [$cm^3\ g^{-1}$]	1.99	1.82	1.68	1.82	2.02
Low-temperature peak (LT) share [%]	26	43	68	75	86
High-temperature peak (HT) share [%]	74	57	32	25	14

Table 3. Comparison of average Co particle and Co crystallite size in the Co/Nd_2O_3 catalysts, determined through H_2-TPD, XRPD and TEM.

Parameter	$Co(10)/Nd_2O_3$	$Co(19)/Nd_2O_3$	$Co(31)/Nd_2O_3$	$Co(39)/Nd_2O_3$	$Co(50)/Nd_2O_3$
$d_{H2\text{-}TPD}$ [nm]	11	22	37	42	47
d_{STEM} [nm]	27	-	36	-	51
d_{XRPD} [nm]	17	21	21	20	19

The data indicates that the average size of Co particles increases with cobalt loading in Co/Nd_2O_3 catalysts. The size growth is nonlinear, but the deviation from a direct proportionality is marginal. The average particle sizes calculated based on chemisorption data are in good accordance with the sizes measured during STEM-EDX observations. However, they are different from the uniform sizes calculated based on diffraction data, which may be caused by the cobalt particles being polycrystalline. The sole existence of the face-centred cubic phase supports the polycrystallinity of cobalt particles. It is in good agreement with the fact that ca. 20 nm and smaller cobalt crystallites tend to occur predominantly in the cubic phase [41]. Without additional structural stabilisation, little to no hexagonal phase is present. It is because in temperatures above 427 °C (i.e., in temperatures lower than the temperature of the activation or the NH_3 synthesis reaction), cobalt, which naturally tends to prevail in the hexagonal close-packed phase [42], undergoes an allotropic transition into the face-centred cubic structure [43].

Activities of Co/Nd_2O_3 catalysts are presented in Table 4. With the increase of cobalt loading, the average reaction rate per catalyst unit mass increases. However, it is worth noting that the observed increase is not directly proportional. The active phase mass increases 5 times, resulting in a less than 2 times increase in average reaction rate. It means that despite the incremental increase of the catalyst surface area and an increase in the number of active sites, the overall efficiency of their utilisation gradually decreases. It is supported by the activity data indicating that for most of the systems, the surface activity is roughly similar, and TOF averages ca. $0.125\ s^{-1}$ (see Figure 8); thus, the influence of the support on the active phase is relatively constant in the studied Co loading range.

Table 4. The activity of Co/Nd_2O_3 catalysts in ammonia synthesis reaction expressed as the average reaction rate (r_{avg}) and TOF.

Parameter	$Co(10)/Nd_2O_3$	$Co(19)/Nd_2O_3$	$Co(31)/Nd_2O_3$	$Co(39)/Nd_2O_3$	$Co(50)/Nd_2O_3$
r_{avg} [$g_{NH3}\ g_{cat}^{-1}\ h^{-1}$]	1.01	1.39	1.27	1.42	1.82

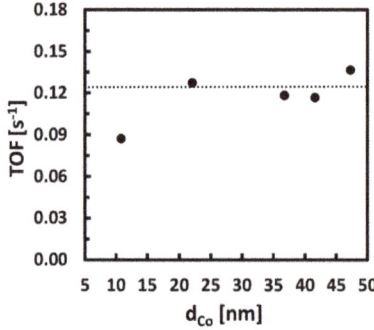

Figure 8. Surface activity (TOF) of Co/Nd_2O_3 catalysts vs Co particle average size. The dotted line represents the average activity level of catalysts. Large dots represent particle sizes calculated on the basis of H_2-TPD results ($d_{H2\text{-}TPD}$, vide Table 3).

However, one may observe that the surface activity of the Co(10)/Nd$_2$O$_3$ catalyst is lower than the average TOF displayed by the others. This state is depicted in Figure 8. It may be related to the smaller size of cobalt particles in this system. Recent studies of cobalt systems deposited on active carbon indicated the optimal cobalt particle size range (20–30 nm) provided the highest activity in ammonia synthesis [27]. Studies of cobalt systems deposited on mixed MgO–La$_2$O$_3$ oxides also showed that cobalt particles of 20 nm of average size yield the highest reaction rate [28].

Our results indicate that when the size of the active phase (metallic cobalt) particles decreases in a Co/Nd$_2$O$_3$ system, the strength of hydrogen binding by the surface increases (Figure 9) as the high-energy to low-energy binding sites ratio grows exponentially. It may seem that when the size of cobalt particles decreases below a certain critical value, i.e., of 20 nm, the above ratio exceeds 1.5 due to the predominance of sites binding hydrogen strongly on the catalyst surface. With the decrease in cobalt particle sizes, more undercoordinated structures, such as close-packed terraces, steps, kinks etc., occur at the cost of open surfaces [41]. These structures bind hydrogen more strongly than flat surfaces [44,45]. Under these circumstances, hydrogen poisoning of the active phase may occur, limiting the activity of the catalyst. Strongly-bound hydrogen blocks the active sites, preventing nitrogen adsorption [14,46–49] and thus inhibiting the rate-determining step of the process [50–53].

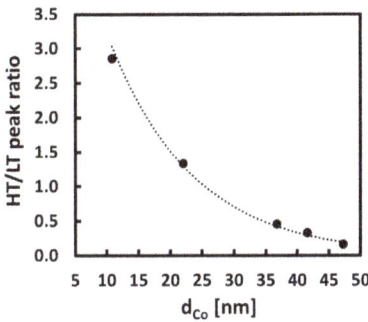

Figure 9. Correlation between the high- and low-temperature H$_2$ desorption peak ratio and Co particle average size of the Co/Nd$_2$O$_3$ catalysts. Large dots represent particle sizes calculated on the basis of H$_2$-TPD results (d$_{H2\text{-}TPD}$, vide Table 3).

3. Materials and Methods

3.1. Catalyst Preparation

The first stage of the catalyst preparation was the calcination of the Nd$_2$O$_3$ support (UMSC Lublin, Poland, 99.99% purity) in the air at 800 °C for 16 h to purify the material by decomposing any neodymium hydroxides and carbonates present [54–56]. Cobalt catalyst precursors were synthesised by the recurrent deposition-precipitation method. The target cobalt content in the catalyst was set in the range of 10 to 50 wt.% at 10 wt.% increments. The systems were labelled as Co(X)/Nd$_2$O$_3$, where X represents the actual cobalt content in the active form of the catalyst. Cobalt(II) carbonate was precipitated with K$_2$CO$_3$ (Avantor Performance Materials Poland S.A., Gliwice, Poland) from an aqueous solution of (Co(NO$_3$)$_2$·6H$_2$O (Acros Organics, Thermo Fischer Scientific, Kandel, Germany) onto the Nd$_2$O$_3$ suspension. The synthesis was conducted at a double molar excess of the precipitating reagent relative to the amount of cobalt nitrate salt. The solution temperature was 85 °C and a mixing speed of 500 rpm was used. The precipitation process was continued until pH = 9 was established, and then the mixture was aged for 1 h. After ageing, the solution was cooled to room temperature and filtered at a reduced pressure (p = 50 mbar). Then the precipitate was washed with distilled water from the residues of potassium, nitrate and carbonate ions to a pH $\approx$ 7 of the filtrate. The purified sludge was dried for 24 h at 120 °C. Obtained materials were calcined in air at 500 °C for 5 h. A neodymium oxide suspension was used as a substrate only for the synthesis of the first Co(10)/Nd$_2$O$_3$ precursor system. For the subsequent Co(X)/Nd$_2$O$_3$ systems, the precursor form of the Co(X-10)/Nd$_2$O$_3$ system was used as the suspension on which another portion of cobalt(II) carbonate was deposited. The cobalt carbonate deposition process was repeated in a precipitation-calcination cycle, increasing the cobalt content by the assumed value of 10 wt.%. The precursors were then tabletted, crushed and sieved to obtain a grain fraction of 0.2–0.63 mm. Fractioned precursors were later subjected to the in-situ activation (reduction) procedure directly before the measurements that required a reduced form of the catalyst. The composition of the obtained catalysts in the reduced (active) form, i.e., metallic cobalt deposited on neodymium oxide, is presented in Table 5.

Table 5. Composition of reduced cobalt catalysts determined with ICP-AES.

Parameter	Co(10)/Nd$_2$O$_3$	Co(19)/Nd$_2$O$_3$	Co(31)/Nd$_2$O$_3$	Co(39)/Nd$_2$O$_3$	Co(50)/Nd$_2$O$_3$
Co content [wt.%]	9.9	19.4	30.8	38.6	49.6
Nd content [wt.%]	77.2	65.6	59.6	46.8	39.5

3.2. Characterisation Methods

Textural parameters of the catalyst precursors were determined by N$_2$ physisorption (Micromeritics Instrument Co., Norcross, GA, USA). Before the measurement, each sample of ca. 0.5 g was evacuated under vacuum $p < 100$ µmHg for 1 h at 90 °C and then for the next 4 h at 300 °C. The specific surface area (S$_{BET}$) of all the materials was determined based on a five-point measurement in the $p/p_0 = 0.01$–0.3 relative pressure range approximated with the Brunauer–Emmett–Teller (BET) isotherm. The total pore volumes (V$_{por}$) were determined based on the multipoint measurement in the $p/p_0 = 0.01$–1 relative pressure range approximated with the Barrett-Joyner-Halenda (BJH) isotherm and were defined for $p/p_0 = 0.995$.

Temperature-programmed hydrogen desorption (H$_2$-TPD) was carried out with the AutoChem 2920 (Micromeritics Instrument Co., Norcross, GA, USA) equipped with a TCD detector, using U-shaped quartz reactors and utilising high purity (≥ 6 N) gases. H$_2$-TPD profiles were captured for reduced catalyst samples. The reduction was carried out in-situ for 16 h at 550 °C in the H$_2$ flow (40 cm^3 min^{-1}). After the reduction process, the samples were rinsed with an inert gas at 620 °C for 1 h and then cooled to 0°C. Hydrogen was adsorbed during cooling from 150 °C to 0 °C and for 15 min at 0 °C. Then the samples were rinsed with Ar for 60 min to eliminate physisorbed H$_2$. Next, hydrogen desorption was conducted during a temperature increase from 0 °C to 800 °C at a heating rate of 10 °C min^{-1}. The concentration of the desorbed hydrogen was measured with a TCD. The desorption profiles were used to calculate the amount of hydrogen desorbed from the metallic Co surface and to estimate the active phase average particle size (assuming the stoichiometry of H$_2$ adsorption on cobalt H:Co = 1:1) [57,58].

The X-ray powder diffraction method (XRPD) determined the phase of precursors and catalysts in the reduced form. The diffraction measurements of catalyst precursors and their active forms were performed in-situ on an X'Pert PRO MPD diffractometer (Philips PANalytical, Almelo, The Netherlands) using CoK$_\alpha$ radiation, operating in a Bragg-Brentano configuration, coupled to an XRK 900reaction chamber (Anton Paar, Graz, Austria). The diffraction data were collected in the 2θ scattering range of 20°–90°, with 0.02° step size and 400 s count time per step. Crystalline phase identification was performed using the PANalytical High Score Plus software with the ICDD PDF4+ 2021 database. The weight concentrations of individual crystalline phases were determined based on full-range refinement of the diffraction profile using the Rietveld method. The catalysts were reduced in an H$_2$ (N5.0, Messer, Warszawie, Poland) stream, with a flow rate of 100 cm^3·min^{-1}, at a temperature of 500 °C for 10 h. Diffraction analyses of both catalyst precursors and their active forms were carried out at an ambient temperature.

The X-ray photoelectron spectroscopy (XPS) analysis was carried out for the oxide precursors and the reduced samples. The measurements were conducted using Al Kα (hν = 1486.6 eV) radiation in a Prevac (Rogów, Poland) system equipped with Scienta SES 2002 electron energy analyser operating at constant transmission energy (Ep = 50 eV). The reduction of precursors was conducted in a High-Pressure Cell (HPC) of an ultra-high vacuum (UHV) system. A small tablet of a sample, approximately 10 mm in diameter, was placed on a sample holder and introduced into HPC. Hydrogen (N5.0 Messer, Poland) was passed through the sample at a constant flow of 20 $cm^3 \cdot min^{-1}$. The sample was heated to 500 °C. The reduction was carried out for 5 h. The sample was then transferred under UHV to the analysis chamber of the electron spectrometer.

TEM investigation of specimens of spent catalysts was carried out using an FEI Titan Cubed 80-300 (FEI Technologies Inc., Hillsboro, OR, USA) microscope operating at 300 kV with a point resolution of 70 pm. The overview images were registered in bright-field TEM mode at magnifications ranging from 7100× to 31,000×, while the HRTEM images were collected at magnifications from 340,000× to 520,000×. The Gatan BM-Ultrascan CCD (Gatan Inc., Pleasanton, CA, USA) camera was used to record both types of images. STEM images were recorded using a HAADF detector and the elemental mapping of the samples was obtained using the in-situ EDX spectrometer. The test material (0.5 mg) was suspended in ethanol (2 mL) and sonicated for 60 s. A drop (20 μL) of the suspension was deposited on a standard TEM copper grid with a diameter of 3 mm, coated with a 30 nm thick amorphous carbon film (EM Resolutions Ltd., Sheffield, UK). After the complete evaporation of the alcohol, the grid was ready for microscopic observations.

Catalytic activity measurements in the ammonia synthesis reaction of Co/Nd_2O_3 catalysts were carried out in a tubular flow reactor supplied with a very pure (99.99995 vol.%) H_2/N_2 = 3 stoichiometric mixture (gas flow rate 70 $dm^3\ h^{-1}$) under semi-industrial conditions: temperature of 470 °C and pressure of 6.3 MPa. A detailed description of the apparatus used can be found elsewhere [15,59]. Before the measurements, the catalyst samples were activated under atmospheric pressure in the reacting gas mixture consecutively at 470 °C for 72 h, then 520 °C for 24 h and 550 °C for 48 h. The outlet concentration of ammonia was measured using an interferometer. The catalytic activity was determined and expressed as an average reaction rate based on the measurement results. Moreover, based on the chemisorption data and activity measurements, an activity of the catalyst's surface (expressed as TOF value) was calculated.

4. Conclusions

In summary, a series of Nd_2O_3-supported Co catalysts of various active phase (metallic cobalt) loading (10–50 wt.%) were synthesised with recurrent deposition precipitation, characterised and tested in NH_3 synthesis. The increase in productivity with the increase in cobalt content was observed in the entire loading range. The increase was linear, however it was disproportionately smaller than the increase in the active phase content. Despite this, catalyst TOF was relatively constant and oscillated around ca. 0.125 s^{-1} in the broad Co loading range. Only the catalyst with the lowest cobalt content displayed a significantly lower TOF. It was attributed to the decrease of the average cobalt particle size below the optimum of 20 nm. The decrease in size entailed the exponential growth of uncoordinated structures on the particle surface composed of strong hydrogen-binding sites, as the cobalt particles are composed of a face-centred cubic phase regardless of the size. The surface state of high-energy binding sites predominance leads to the poisoning of the cobalt surface with hydrogen under the ammonia synthesis reaction conditions. Hence, the activity is limited by blocking of the active sites for nitrogen adsorption, thus inhibiting the rate-determining step of the process.

Author Contributions: Conceptualization, W.P., M.Z. and W.R.-P.; methodology, W.P, M.Z. and W.R.-P.; investigation, W.P., A.F., A.A., D.M., P.D., M.Z., H.R., G.G. and W.R.-P.; writing—original draft preparation, W.P., M.Z.; writing—review and editing, W.P., M.Z. and W.R.-P.; visualisation, W.P.; supervision, W.P., M.Z. and W.R.-P. All authors have read and agreed to the published version of the manuscript.

Funding: This research received no external funding.

Data Availability Statement: All data is available within the paper.

Conflicts of Interest: The authors declare no conflict of interest.

References

1. Smith, C.; Hill, A.K.; Torrente-Murciano, L. Current and future role of Haber–Bosch ammonia in a carbon-free energy landscape. *Energy Environ. Sci.* **2020**, *13*, 331–344. [CrossRef]
2. Jennings, J.R. *Catalytic Ammonia Synthesis—Fundamentals and Practice*; Springer: New York, NY, USA, 2008.
3. Liu, H. *Ammonia Synthesis Catalysts*; World Scientific: Singapore, 2013.
4. Faria, J.A. Renaissance of ammonia synthesis for sustainable production of energy and fertilizers. *Curr. Opin. Green Sustain. Chem.* **2021**, *29*, 100466. [CrossRef]
5. Nilsson, A.; Pettersson, L.G.M.; Hammer, B.; Bligaard, T.; Christensen, C.H.; Nørskov, J.K. The electronic structure effect in heterogeneous catalysis. *Catal. Lett.* **2005**, *100*, 111–114. [CrossRef]
6. Hagen, S.; Barfod, R.; Fehrmann, R.; Jacobsen, C.J.H.H.; Teunissen, H.T.; Ståhl, K.; Chorkendorff, I. New efficient catalyst for ammonia synthesis: Barium-promoted cobalt on carbon. *Chem. Commun.* **2002**, *11*, 1206–1207. [CrossRef]
7. Hagen, S.; Barfod, R.; Fehrmann, R.; Jacobsen, C.J.H.; Teunissen, H.T.; Chorkendorff, I. Ammonia synthesis with barium-promoted iron-cobalt alloys supported on carbon. *J. Catal.* **2003**, *214*, 327–335. [CrossRef]
8. Raróg-Pilecka, W.; Miśkiewicz, E.; Kępiński, L.; Kaszkur, Z.; Kielar, K.; Kowalczyk, Z. Ammonia synthesis over barium-promoted cobalt catalysts supported on graphitised carbon. *J. Catal.* **2007**, *249*, 24–33. [CrossRef]
9. Karolewska, M.; Truszkiewicz, E.; Wściseł, M.; Mierzwa, B.; Kępiński, L.; Raróg-Pilecka, W. Ammonia synthesis over a Ba and Ce-promoted carbon-supported cobalt catalyst. Effect of the cerium addition and preparation procedure. *J. Catal.* **2013**, *303*, 130–134. [CrossRef]
10. Lin, B.; Qi, Y.; Wei, K.; Lin, J. Effect of pretreatment on ceria-supported cobalt catalyst for ammonia synthesis. *RSC Adv.* **2014**, *4*, 38093. [CrossRef]
11. Lin, B.; Liu, Y.; Heng, L.; Ni, J.; Lin, J.; Jiang, L. Effect of ceria morphology on the catalytic activity of Co/CeO$_2$ catalyst for ammonia synthesis. *Catal. Commun.* **2017**, *101*, 15–19. [CrossRef]
12. Lin, B.; Liu, Y.; Heng, L.; Ni, J.; Lin, J.; Jiang, L. Effect of barium and potassium promoter on Co/CeO$_2$ catalysts in ammonia synthesis. *J. Rare Earths* **2018**, *36*, 703–707. [CrossRef]
13. Sato, K.; Miyahara, S.; Tsujimaru, K.; Wada, Y.; Toriyama, T.; Yamamoto, T.; Matsumura, S.; Inazu, K.; Mohri, H.; Iwasa, T.; et al. Barium Oxide Encapsulating Cobalt Nanoparticles Supported on Magnesium Oxide: Active Non-Noble Metal Catalysts for Ammonia Synthesis under Mild Reaction Conditions. *ACS Catal.* **2021**, *11*, 13050–13061. [CrossRef]
14. Ronduda, H.; Zybert, M.; Patkowski, W.; Tarka, A.; Jodłowski, P.; Kępiński, L.; Sarnecki, A.; Moszyński, D.; Raróg-Pilecka, W. Tuning the catalytic performance of Co/Mg-La system for ammonia synthesis via the active phase precursor introduction method. *Appl. Catal. A Gen.* **2020**, *598*, 117553. [CrossRef]
15. Ronduda, H.; Zybert, M.; Patkowski, W.; Tarka, A.; Ostrowski, A.; Raróg-Pilecka, W. Kinetic studies of ammonia synthesis over a barium-promoted cobalt catalyst supported on magnesium–lanthanum mixed oxide. *J. Taiwan Inst. Chem. Eng.* **2020**, *114*, 241–248. [CrossRef]
16. Ronduda, H.; Zybert, M.; Patkowski, W.; Ostrowski, A.; Jodłowski, P.; Szymański, D.; Kępiński, L.; Raróg-Pilecka, W. Boosting the Catalytic Performance of Co/Mg/La Catalyst for Ammonia Synthesis by Selecting a Pre-Treatment Method. *Catalysts* **2021**, *11*, 941. [CrossRef]
17. Ronduda, H.; Zybert, M.; Patkowski, W.; Ostrowski, A.; Jodłowski, P.; Szymański, D.; Kępiński, L.; Raróg-Pilecka, W. A high performance barium-promoted cobalt catalyst supported on magnesium–lanthanum mixed oxide for ammonia synthesis. *RSC Adv.* **2021**, *11*, 14218–14228. [CrossRef]
18. Ronduda, H.; Zybert, M.; Patkowski, W.; Ostrowski, A.; Jodłowski, P.; Szymański, D.; Kępiński, L.; Raróg-Pilecka, W. Development of cobalt catalyst supported on MgO–Ln$_2$O$_3$ (Ln = La, Nd, Eu) mixed oxide systems for ammonia synthesis. *Int. J. Hydrog. Energy* **2022**, *47*, 6666–6678. [CrossRef]
19. Inoue, Y.; Kitano, M.; Tokunari, M.; Taniguchi, T.; Ooya, K.; Abe, H.; Niwa, Y.; Sasase, M.; Hara, M.; Hosono, H. Direct Activation of Cobalt Catalyst by 12CaO$_7$Al$_2$O$_3$ Electride for Ammonia Synthesis. *ACS Catal.* **2019**, *9*, 1670–1679. [CrossRef]
20. Gao, W.; Wang, P.; Guo, J.; Chang, F.; He, T.; Wang, Q.; Wu, G.; Chen, P. Barium Hydride-Mediated Nitrogen Transfer and Hydrogenation for Ammonia Synthesis: A Case Study of Cobalt. *ACS Catal.* **2017**, *7*, 3654–3661. [CrossRef]
21. Raróg-Pilecka, W.; Karolewska, M.; Truszkiewicz, E.E.; Iwanek, E.; Mierzwa, B. Cobalt Catalyst Doped with Cerium and Barium Obtained by Co-Precipitation Method for Ammonia Synthesis Process. *Catal. Lett.* **2011**, *141*, 678–684. [CrossRef]

22. Tarka, A.; Patkowski, W.; Zybert, M.; Ronduda, H.; Wieciński, P.; Adamski, P.; Sarnecki, A.; Moszyński, D.; Raróg-Pilecka, W. Synergistic interaction of cerium and barium-new insight into the promotion effect in cobalt systems for ammonia synthesis. *Catalysts* **2020**, *10*, 658. [CrossRef]
23. Zybert, M.; Wyszyńska, M.; Tarka, A.; Patkowski, W.; Ronduda, H.; Mierzwa, B.; Kępiński, L.; Sarnecki, A.; Moszyński, D.; Raróg-Pilecka, W. Surface enrichment phenomenon in the Ba-doped cobalt catalyst for ammonia synthesis. *Vacuum* **2019**, *168*, 108831. [CrossRef]
24. Patkowski, W.; Kowalik, P.; Antoniak-Jurak, K.; Zybert, M.; Ronduda, H.; Mierzwa, B.; Próchniak, W.; Raróg-Pilecka, W. On the Effect of Flash Calcination Method on the Characteristics of Cobalt Catalysts for Ammonia Synthesis Process. *Eur. J. Inorg. Chem.* **2021**, *2021*, 1518–1529. [CrossRef]
25. Karolewska, M.; Truszkiewicz, E.; Mierzwa, B.; Keopiński, L.; Raróg-Pilecka, W. Ammonia synthesis over cobalt catalysts doped with cerium and barium. Effect of the ceria loading. *Appl. Catal. A Gen.* **2012**, *445–446*, 280–286. [CrossRef]
26. Zybert, M.; Tarka, A.; Mierzwa, B.; Kępiński, L.; Raróg-Pilecka, W. Promotion effect of lanthanum on the Co/La/Ba ammonia synthesis catalysts—The influence of lanthanum content. *Appl. Catal. A Gen.* **2016**, *515*, 16–24. [CrossRef]
27. Zybert, M.; Tarka, A.; Patkowski, W.; Ronduda, H.; Mierzwa, B.; Kępiński, L.; Raróg-Pilecka, W. Structure Sensitivity of Ammonia Synthesis on Cobalt: Effect of the Cobalt Particle Size on the Activity of Promoted Cobalt Catalysts Supported on Carbon. *Catalysts* **2022**, *12*, 1285. [CrossRef]
28. Ronduda, H.; Zybert, M.; Patkowski, W.; Sobczak, K.; Moszyński, D.; Albrecht, A.; Sarnecki, A.; Raróg-Pilecka, W. On the effect of metal loading on the performance of Co catalysts supported on mixed MgO–La$_2$O$_3$ oxides for ammonia synthesis. *RSC Adv.* **2022**, *12*, 33876–33888. [CrossRef]
29. Niwa, Y.; Aika, K. The Effect of Lanthanide Oxides as a Support for Ruthenium Catalysts in Ammonia Synthesis. *J. Catal.* **1996**, *162*, 138–142. [CrossRef]
30. Miyahara, S.I.; Sato, K.; Kawano, Y.; Imamura, K.; Ogura, Y.; Tsujimaru, K.; Nagaoka, K. Ammonia synthesis over lanthanoid oxide–supported ruthenium catalysts. *Catal. Today* **2020**, *376*, 2–5. [CrossRef]
31. Sato, K.; Imamura, K.; Kawano, Y.; Miyahara, S.; Yamamoto, T.; Matsumura, S.; Nagaoka, K. A low-crystalline ruthenium nano-layer supported on praseodymium oxide as an active catalyst for ammonia synthesis. *Chem. Sci.* **2017**, *8*, 674–679. [CrossRef]
32. Imamura, K.; Miyahara, S.I.; Kawano, Y.; Sato, K.; Nakasaka, Y.; Nagaoka, K. Kinetics of ammonia synthesis over Ru/Pr2O3. *J. Taiwan Inst. Chem. Eng.* **2019**, *105*, 50–56. [CrossRef]
33. Bernal, S.; Botana, F.J.; García, R.; Rodríguez-Izquierdo, J.M. Behaviour of rare earth sesquioxides exposed to atmospheric carbon dioxide and water. *React. Solids* **1987**, *4*, 23–40. [CrossRef]
34. Bernal, S.; Botana, F.J.; García, R.; Rodríguez-Izquierdo, J.M. Study of the interaction of two hexagonal neodymium oxides with atmospheric CO_2 and H_2O. *J. Mater. Sci.* **1988**, *23*, 1474–1480. [CrossRef]
35. Turcotte, R.P.; Sawyer, J.O.; Eyring, L. Rare earth dioxymonocarbonates and their decomposition. *Inorg. Chem.* **1969**, *8*, 238–246. [CrossRef]
36. Biesinger, M.C.; Payne, B.P.; Grosvenor, A.P.; Lau, L.W.M.; Gerson, A.R.; Smart, R.S.C. Resolving surface chemical states in XPS analysis of first row transition metals, oxides and hydroxides: Cr, Mn, Fe, Co and Ni. *Appl. Surf. Sci.* **2011**, *257*, 2717–2730. [CrossRef]
37. Garces, L.J.; Hincapie, B.; Zerger, R.; Suib, S.L. The Effect of Temperature and Support on the Reduction of Cobalt Oxide: An in Situ X-ray Diffraction Study. *J. Phys. Chem. C* **2015**, *119*, 5484–5490. [CrossRef]
38. Zybert, M.; Karasińska, M.; Truszkiewicz, E.; Mierzwa, B.; Raróg-Pilecka, W. Properties and activity of the cobalt catalysts for NH 3 synthesis obtained by co-precipitation—the effect of lanthanum addition. *Pol. J. Chem. Technol.* **2015**, *17*, 138–143. [CrossRef]
39. Ciufo, R.A.; Han, S.; Floto, M.E.; Eichler, J.E.; Henkelman, G.; Mullins, C.B. Hydrogen desorption from the surface and subsurface of cobalt. *Phys. Chem. Chem. Phys.* **2020**, *22*, 15281–15287. [CrossRef]
40. Huesges, Z.; Christmann, K. Interaction of Hydrogen with a Cobalt(0001) Surface. *Z. Für Phys. Chem.* **2013**, *227*, 881–899. [CrossRef]
41. Kitakami, O.; Sato, H.; Shimada, Y.; Sato, F.; Tanaka, M. Size effect on the crystal phase of cobalt fine particles. *Phys. Rev. B* **1997**, *56*, 13849–13854. [CrossRef]
42. Liu, J.X.; Li, W.X. Theoretical study of crystal phase effect in heterogeneous catalysis. *Wiley Interdiscip. Rev. Comput. Mol. Sci.* **2016**, *6*, 571–583. [CrossRef]
43. Freels, M.; Liaw, P.K.; Jiang, L.; Klarstrom, D.L. Advanced Structural Materials: Properties, Design Optimization, and Applications. In *Advanced Structural Materials: Properties, Design Optimization, and Applications*; Soboyejo, W.O., Srivatsan, T.S., Eds.; CRC Press: Boca Raton, FL, USA, 2007; pp. 187–224.
44. Weststrate, C.J.; Mahmoodinia, M.; Farstad, M.H.; Svenum, I.-H.; Strømsheim, M.D.; Niemantsverdriet, J.W.; Venvik, H.J. Interaction of hydrogen with flat (0001) and corrugated (11–20) and (10–12) cobalt surfaces: Insights from experiment and theory. *Catal. Today* **2020**, *342*, 124–130. [CrossRef]
45. Weststrate, C.J.; Garcia Rodriguez, D.; Sharma, D.; Niemantsverdriet, J.W. Structure-dependent adsorption and desorption of hydrogen on FCC and HCP cobalt surfaces. *J. Catal.* **2022**, *405*, 303–312. [CrossRef]
46. Fernández, C.; Bion, N.; Gaigneaux, E.M.; Duprez, D.; Ruiz, P. Kinetics of hydrogen adsorption and mobility on Ru nanoparticles supported on alumina: Effects on the catalytic mechanism of ammonia synthesis. *J. Catal.* **2016**, *344*, 16–28. [CrossRef]
47. Kojima, R.; Aika, K. Cobalt molybdenum bimetallic nitride catalysts for ammonia synthesis. *Appl. Catal. A Gen.* **2001**, *218*, 121–128. [CrossRef]

48. Rosowski, F.; Hornung, A.; Hinrichsen, O.; Herein, D.; Muhler, M.; Ertl, G. Ruthenium catalysts for ammonia synthesis at high pressures: Preparation, characterization, and power-law kinetics. *Appl. Catal. A Gen.* **1997**, *151*, 443–460. [CrossRef]
49. Rivera Rocabado, D.S.; Aizawa, M.; Noguchi, T.G.; Yamauchi, M.; Ishimoto, T. Uncovering the Mechanism of the Hydrogen Poisoning on Ru Nanoparticles via Density Functional Theory Calculations. *Catalysts* **2022**, *12*, 331. [CrossRef]
50. Boudart, M. Kinetics and mechanism of ammonia synthesis. *Catal. Rev. Eng.* **1981**, *23*, 1–15. [CrossRef]
51. Ertl, G.; Lee, S.B.; Weiss, M. Adsorption of nitrogen on potassium promoted Fe(111) and (100) surfaces. *Surf. Sci.* **1982**, *114*, 527–545. [CrossRef]
52. Bozso, F.; Ertl, G.; Grunze, M.; Weiss, M. Chemisorption of hydrogen on iron surfaces. *Appl. Surf. Sci.* **1977**, *1*, 103–119. [CrossRef]
53. Ertl, G. Surface Science and Catalysis—Studies on the Mechanism of Ammonia Synthesis: The P. H. Emmett Award Address. *Catal. Rev.* **1980**, *21*, 201–223. [CrossRef]
54. Rosynek, M.P. Catalytic Properties of Rare Earth Oxides. *Catal. Rev.—Sci. Eng.* **1977**, *16*, 111–154. [CrossRef]
55. Alvero, R.; Odriozola, J.A.; Trillo, J.M.; Bernal, S. Lanthanide oxides: Preparation and ageing. *J. Chem. Soc. Dalt. Trans.* **1984**, *2*, 87. [CrossRef]
56. Bernal, S.; Blanco, G.; Calvino, J.J.; Omil, J.A.P.; Pintado, J.M. Some major aspects of the chemical behavior of rare earth oxides: An overview. *J. Alloy. Compd.* **2006**, *408–412*, 496–502. [CrossRef]
57. Reuel, R.C.; Bartholomew, C.H. The stoichiometries of H2 and CO adsorptions on cobalt: Effects of support and preparation. *J. Catal.* **1984**, *85*, 63–77. [CrossRef]
58. Borodziński, A.; Bonarowska, M. Relation between Crystallite Size and Dispersion on Supported Metal Catalysts. *Langmuir* **1997**, *13*, 5613–5620. [CrossRef]
59. Kowalczyk, Z. Effect of potassium on the high pressure kinetics of ammonia synthesis over fused iron catalyst. *Catal. Lett.* **1996**, *37*, 173–179. [CrossRef]

Disclaimer/Publisher's Note: The statements, opinions and data contained in all publications are solely those of the individual author(s) and contributor(s) and not of MDPI and/or the editor(s). MDPI and/or the editor(s) disclaim responsibility for any injury to people or property resulting from any ideas, methods, instructions or products referred to in the content.

Article

Structure Sensitivity of Ammonia Synthesis on Cobalt: Effect of the Cobalt Particle Size on the Activity of Promoted Cobalt Catalysts Supported on Carbon

Magdalena Zybert [1,*], Aleksandra Tarka [1], Wojciech Patkowski [1], Hubert Ronduda [1], Bogusław Mierzwa [2], Leszek Kępiński [3] and Wioletta Raróg-Pilecka [1,*]

1. Faculty of Chemistry, Warsaw University of Technology, Noakowskiego 3, 00-664 Warsaw, Poland
2. Institute of Physical Chemistry, Polish Academy of Sciences, Kasprzaka 44/52, 01-224 Warsaw, Poland
3. Institute of Low Temperature and Structure Research, Polish Academy of Sciences, Okólna 2, 50-950 Wrocław, Poland
* Correspondence: magdalena.zybert@pw.edu.pl (M.Z.); wioletta.pilecka@pw.edu.pl (W.R.-P.)

Abstract: This work presents a size effect, i.e., catalyst surface activity, as a function of active phase particle size in a cobalt catalyst for ammonia synthesis. A series of cobalt catalysts supported on carbon and doped with barium was prepared, characterized (TEM, XRPD, and H_2 chemisorption), and tested in ammonia synthesis (9.0 MPa, 400 °C, H_2/N_2 = 3, 8.5 mol% of NH_3). The active phase particle size was varied from 3 to 45 nm by changing the metal loading in the range of 4.9–67.7 wt%. The dependence of the reaction rate expressed as TOF on the active phase particle size revealed an optimal size of cobalt particles (20–30 nm), ensuring the highest activity of the cobalt catalyst in the ammonia synthesis reaction. This indicated that the ammonia synthesis reaction on cobalt is a structure-sensitive reaction. The observed effect may be attributed to changes in the crystalline structure, i.e., the appearance of the hcp Co phase for the particles with a diameter of 20–30 nm.

Keywords: structure sensitivity; size effect; cobalt catalyst; ammonia synthesis; carbon support

1. Introduction

The structure sensitivity of a reaction [1–4] is a concept established in the 1960s when Boudart noticed that among the reactions carried out in the presence of heterogeneous catalysts, there are those where the rate changes with the size (diameter) of the active phase particles [1]. Van Santen also reported that an increase in the size of metal particles may cause changes in the reaction rate expressed as TOF (turnover frequency), i.e., the activity of a single active site of a catalyst [4]. It was also pointed out that the reaction rate may increase or reach its maximum and then decrease with increasing particle size. However, when considering a size effect, i.e., TOF as a function of metal particle size, it should be underlined that the surface structure may also change with changes in particle size. The popular terrace-ledge-kink (TLK) model assumes the heterogeneity of a surface structure connected with the presence of steps, kinks, adatoms, and vacancies. In the case of very small metal particles or their larger clusters, it is more appropriate to describe the surface structure using the coordination model, where the atoms present on the surface differ from the atoms inside the particle by having an incomplete set of adjacent atoms. Therefore, surface atoms are described by a coordination number, i.e., the number of nearest adjacent atoms (e.g., the C_i atom has i other atoms adjacent to it, while the B_j active site has j closest neighbors). When the particle size of the catalyst's active phase changes, the surface structure also differs because the relative concentration of the surface atoms and active sites changes. Their number increases with the increasing crystallite size. Some studies have shown that the presence of atoms or active sites with specific coordination can be attributed to the high activity of the catalyst [5,6]. Different particle sizes may also favor the formation of various crystallographic structures of the active phase [7].

The catalytic reaction of hydrogen and nitrogen in the presence of iron, leading to the formation of ammonia, is a prime example of a reaction highly dependent on the structure of the active phase [5,8–16]. Ertl et al. showed that dissociative nitrogen adsorption, i.e., the rate-determining step of the NH_3 synthesis reaction, strongly depends on the iron surface structure [8,9]. The activity of the investigated iron surfaces in relation to dissociative nitrogen adsorption changes in the order Fe (111) > Fe (100) > Fe (110) [8]. This sequence is consistent with the results obtained by Samoraj et al. [5,10,13], where it was observed that the Fe (111) and Fe (211) surfaces are much more active in the synthesis of ammonia than the Fe (100), Fe (210), and Fe (110). It was found that the differences in the activity of these structures result directly from the arrangement of Fe atoms and, more specifically, from their coordination. The Fe (111) is the best Fe single crystal surface for NH_3 synthesis [16]. The high activity of the Fe (111) surface is related to the presence of C7 atoms, i.e., Fe atoms with seven nearest neighbors [5]. Assuming that the surface structure of iron particles may change with their size, Dumesic et al. investigated the properties of supported iron catalysts (Fe/MgO) with iron particles of different sizes (d = 1.5, 4, 10, 30 nm) [11,12]. The research confirmed that the activity in ammonia synthesis (TOF) increases with the increasing size of iron particles. This effect was explained by a lower number of C7 atoms on the surface of smaller particles [11]. They also noticed that the pretreatment of the iron catalyst of small Fe particles in a nitrogen atmosphere may cause a reconstruction of the iron crystallites surface, and the concentration of C7 atoms may be increased [12]. Theoretical studies of bcc Fe surface using DFT calculation and micro-kinetics analysis [14] revealed that Fe (211), (310), and (221) surfaces, which consist of active C7 and/or B5 sites, dominate the overall reactivity when iron particles are larger than 6 nm. Further decreasing the particle size to 2.3–6 nm, the corresponding TOF is two or three times lower due to the absence of the highly active Fe (221) surface. However, the reactivity is substantially reduced for particles smaller than 2 nm because no active C7 and/or B5 sites are exposed.

Numerous studies also indicate that the ammonia synthesis reaction on the ruthenium catalyst is structure-sensitive [6,17–22]. Dahl et al. [17–19] showed that the N_2 dissociation on the Ru (0001) surface is dominated by steps (at least 9 orders of magnitude higher than on the terraces), which is a result of a combination of electronic and geometrical effects coexisting in the ruthenium catalyst. Jacobsen et al. stated [6] that B5 sites, i.e., a system of five atoms with a specific configuration, are responsible for the high activity of ruthenium. Van Hardeveld and van Montfoort also observed the presence of B5 sites on the platinum, palladium, and nickel surfaces [20], indicating that these sites are responsible for strong nitrogen adsorption. Both the studies of van Hardeveld [20] and Jacobsen [6] showed that the number of B5 sites depends on the size of metal crystallites. In the case of particles smaller than 1 nm, B5 sites' contribution to the metal surface is small. Their number increases with increasing particle size reaching the maximum for particles in the range of 1.8–2.5 nm and gradually decreasing for larger particles [6]. Studies of ruthenium catalysts in the ammonia synthesis reaction revealed that the reaction rate is a function of the Ru crystallite size. Liang showed [21] that TOF increased with an increase in the average Ru crystallite size in the range of 1.7–10.3 nm for the promoted catalyst supported on carbon (Ru-K/C). In the work [22], Raróg-Pilecka et al. also reported that with increasing ruthenium content on the carbon support in the range of 1–32 wt%, the average size of Ru crystallites increased monotonically from 1 nm to 4 nm and along with it, the activity (TOF) increased. It was also noted that extrapolation of the TOF value for crystallites smaller than the examined one suggests that very fine ruthenium crystallites (i.e., smaller than 0.7–0.8 nm) may be completely inactive in the ammonia synthesis reaction (a critical size). Li et al. [23] reported that sub-nanometric Ru catalysts not only exhibit performance different from that of NPs, but also follow a different route for N_2 activation.

Ammonia synthesis is a structure-sensitive reaction when conducted in the presence of a cobalt catalyst. Rambeau noted that cobalt's activity in this reaction strongly depends on its allotropic form [24]. The hexagonal close-packed (hcp) phase has twice the activity (expressed as TOF) of the face-centered cubic (fcc) phase. Density functional theory (DFT) calculations showed that the activity of the hcp Co (10$\bar{1}$2) and hcp Co (11$\bar{2}$1) surfaces in the N_2 dissociation process significantly exceeds the activity of the other surfaces, which is reflected in the high activity of the hcp cobalt phase in the ammonia synthesis reaction [25]. However, the effect of cobalt particle size on the catalytic properties of cobalt catalysts in ammonia synthesis has not been documented so far.

This study aimed to investigate the effect of cobalt particle size on the activity of the cobalt catalyst in the ammonia synthesis reaction. A series of barium-promoted cobalt catalysts supported on graphitized carbon was prepared. As reported in the previous studies of our group [26,27] and other researchers [28,29], this type of catalyst is a promising alternative system for the synthesis of ammonia, exhibiting higher activities at synthesis temperatures than the commercial, multipromoted iron catalyst as well as a lower ammonia inhibition. Co is normally quite inert toward N_2 dissociation, and as a result it shows low activity in ammonia synthesis reactions. Hence, the addition of a barium promoter is required to boost its performance. The promoting effect of barium results not only from its electronic and/or structural influence [30,31], but, according to the latest research [32], is related to the reduction of the spin polarization of the neighboring Co atoms defining the active site for N_2 dissociation. The promoted cobalt systems deposited on a carrier represent a good object for studying the particle-size dependence of activity. In the present studies, the active phase particle size was varied by changing the metal loading in the range of 4.9–67.7 wt%. The characterization of cobalt particle size was carried out using the chemisorption method (H_2-TPD). The sorption experiments were supplemented with XRPD and TEM measurements. An in-depth discussion of the Co particle size–NH_3 catalyst activity relationship was provided, and the optimal Co particle size ensuring the most favorable catalytic properties of the cobalt systems was determined.

2. Results

2.1. Characterization of a Carbon Support

The material used as support was a modified carbon of a well-developed texture. The total surface area and total pore volume were 1611 $m^2\ g^{-1}$ and 1.02 $cm^3\ g^{-1}$, respectively. A significant contribution of micropores was observed, which accounted for 85% of the total surface area and 80% of the total porosity. This characteristic is essential for preparing highly loaded Ba-Co/C catalysts of high metal dispersion [26]. The structural analysis showed that the support exhibited features characteristic of turbostratic carbon [33–36], having structural ordering between the amorphous carbon phase and crystalline graphite phase. In the diffraction pattern depicted in Figure 1a, the broad reflections derived from the (002) and (004) planes and asymmetric (10) signal characteristic of turbostratic carbon were visible. Moreover, a sharp reflex at 2θ = 26.2°, assigned to the presence of a small amount of graphite in the support, was observed. The size of turbostratic stacks in the (002) direction was approximately 1 nm, and the average size of graphite stacks was approximately 6 nm (both values were estimated based on Scherrer's equation). The parallel graphene layers visible in the TEM image (Figure 1b) and the carbon-derived rings in the selected area electron diffraction (SAED) pattern (Figure 1c) confirmed the partial graphitization of the carbon support.

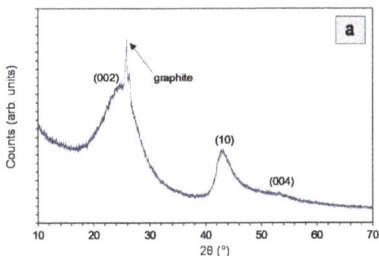
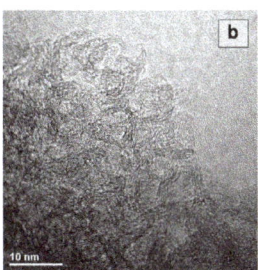
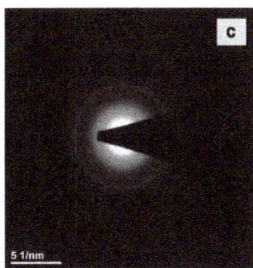

Figure 1. The structure of the carbon support: (**a**) XRPD pattern, (**b**) HRTEM image, (**c**) SAED pattern.

2.2. Characterization of the Active Phase (Cobalt) Particles

Structural characterization (TEM and XRPD) studies were carried out for the Ba-Co/C catalysts previously tested in NH_3 synthesis reaction (i.e., the spent catalysts after exposure to air—ex situ measurements). The diffraction profiles for the catalysts are depicted in Figure 2. The reflections at $2\theta = 44.2°$, $51.5°$, $75.9°$, and $92.2°$ corresponding to the fcc Co phase were identified for all the tested samples. As expected, the contribution of cobalt to the diffraction profiles strongly depended on the metal loading. The intensity of the reflections assigned to the fcc Co phase increased with increasing Co content, which suggested an increase in the crystallinity of this phase. The most intensive reflections corresponded to the catalysts with the highest metal content (Ba-Co43.1/C and Ba-Co67.7/C). For these systems, the fcc Co signals were asymmetric, which can be assigned to stacking faults or the presence of mixed fcc Co and hcp Co phases. Low-intensity signals at $2\theta = 41.7°$ and $47.6°$ can be attributed to the hcp Co phase. Moreover, the reflections at $2\theta = 43°$ and $79°$ assigned to carbon were observed for all the catalysts. No reflections indicating the presence of barium compounds were recorded, suggesting their good dispersion or amorphous form. The average size of cobalt crystallites estimated based on fcc Co reflections (d_{Co-fcc}) is presented in Table 1. The smallest Co crystallites of 14 nm were observed for the sample with the lowest metal content (Ba-Co4.9/C) and the largest crystallites of 25 nm for the catalyst with the highest Co content (Ba-Co67.7/C). For other systems, cobalt crystallites were of similar size (approximately 20 nm).

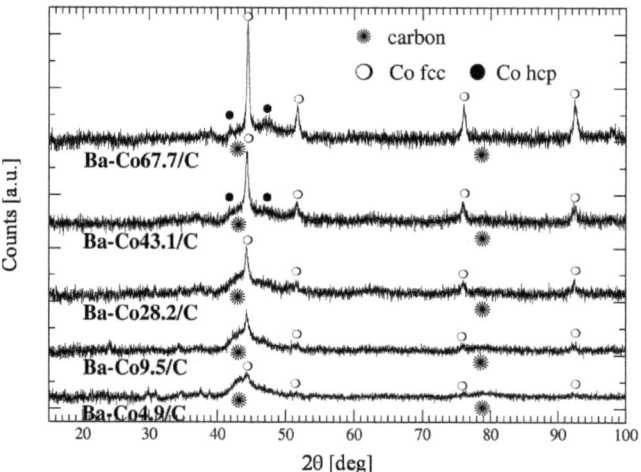

Figure 2. XRPD patterns of the Ba-Co/C catalysts (the spent catalysts after exposure to air—ex situ measurements).

Table 1. Average cobalt crystallite and cobalt particle size in the Ba-Co/C catalysts based on XRPD and TEM measurements (the spent catalysts after exposure to air—ex situ measurements).

Catalyst	d_{Co-fcc} [1] (nm)	$d_{core-shell}$ [2] (nm)	d_{core} [3] (nm)
Ba-Co4.9/C	14	–	–
Ba-Co9.5/C	21	20	15
Ba-Co28.2/C	19	26	19
Ba-Co43.1/C	18	22	16
Ba-Co67.7/C	25	41	31

[1] the average cobalt crystallite size determined based on the Co fcc reflections from the XRPD measurements; [2] the average size of the "core-shell" particles observed by TEM; [3] the average size of the metallic core of the particles observed by TEM.

TEM investigations (Figure 3) revealed that differences in the cobalt particle size were obtained depending on the active phase content. Changes in the size and number of Co particles on the support surface were clearly visible for the catalysts with cobalt content: 4.9 wt% (Figure 3a), 48.1 wt% (Figure 3b) and 67.7 wt% (Figure 3c). The preparation procedure enabled to obtain catalysts with sufficient dispersion (i.e., with a high Co particle density per support surface unit) even when the active phase content was relatively high (e.g., Ba-Co43.1/C, Figure 3b). Moreover, it was observed that in the studied catalysts, the cobalt particles were either partially (on the surface) or fully oxidized. The particles with a diameter smaller than 10 nm were almost entirely oxidized (Figure 4a), whereas the larger particles formed a "core-shell" type structure (Figure 4b) with a core of metallic cobalt surrounded by an oxide layer. Cobalt oxidation is due to exposure of the catalysts to air after their removal from the ammonia synthesis reactor (ex situ TEM measurements). This caused some difficulties in terms of an adequate cobalt particle size estimation during TEM experiments, especially for the catalysts, where small particles (<10 nm) dominated (Ba-Co4.9/C). The contribution of these particles increased with the decrease of the total Co content in the sample. Therefore, the error in the average particle size determination increased for the catalysts of low Co content.

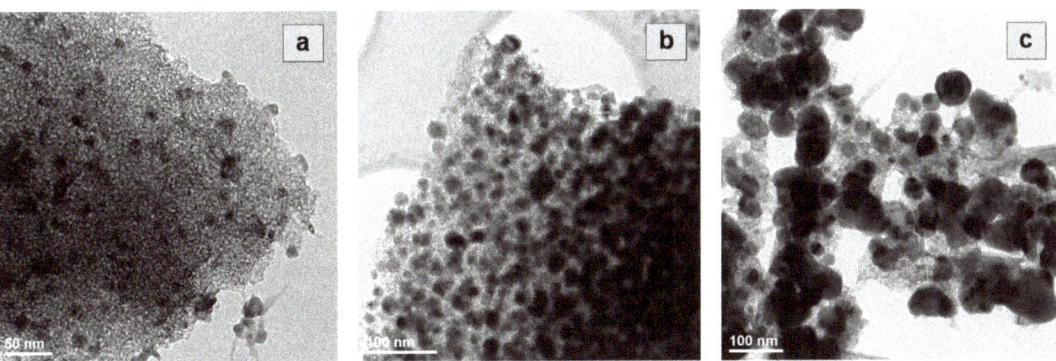

Figure 3. TEM images of the selected catalysts of (**a**) low (Ba-Co4.9/C), (**b**) medium (Ba-Co43.1/C), and (**c**) high (Ba-Co67.7/C) cobalt content (the spent catalysts after exposure to air—ex situ measurements).

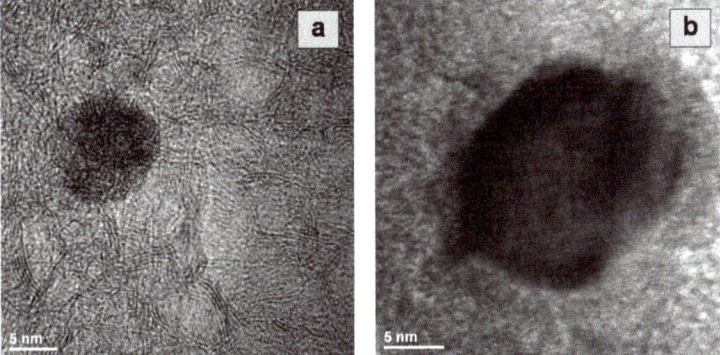

Figure 4. HRTEM images of a single Co particle: (**a**) entirely oxidized—Ba-Co4.9/C catalyst, (**b**) surrounded by an oxide layer—"core-shell" structure of Ba-Co28.2/C catalyst (the spent catalysts after exposure to air—ex situ measurements).

The average size of core-shell particles ($d_{core-shell}$) and the average size of a metallic core (d_{core}) estimated based on TEM results are presented in Table 1. The average size of the metallic core was the largest for the catalyst with the highest Co content (Ba-Co67.7/C) and exceeded 30 nm. For other systems, d_{core} values were similar. The average size of the metallic core was in good agreement with the average cobalt crystallite size (d_{Co-fcc}) determined by XRPD (Table 1).

Figure 5 depicts the profiles of hydrogen desorption from the surface of the selected catalysts of low (Ba-Co4.9/C), medium (Ba-Co43.1/C), and high (Ba-Co67.7/C) cobalt content. Two signals of similar characteristics but different intensities were observed on the profiles of all the studied catalysts. The low-temperature signal with a maximum at ca. 100 °C corresponded to hydrogen desorption from weakly binding adsorption sites. The high-temperature signal with a maximum of 410–500 °C was assigned to the presence of strongly binding hydrogen adsorption sites. However, one can see from Figure 5 that, depending on the cobalt content in the catalyst, adsorption sites of different natures dominated on the surface of the active phase. In the catalyst of low cobalt content (Ba-Co (4.9)/C), the similar intensity of both signals indicated the presence of a comparable number of adsorption sites weakly and strongly binding hydrogen with a slight predominance of weak adsorption sites. In the case of the Ba-Co (43.1)/C catalyst, adsorption sites of low hydrogen binding strength were more pronounced. Strong hydrogen binding sites clearly dominated on the surface of the catalyst with the highest active phase content (Ba-Co (67.7)/C).

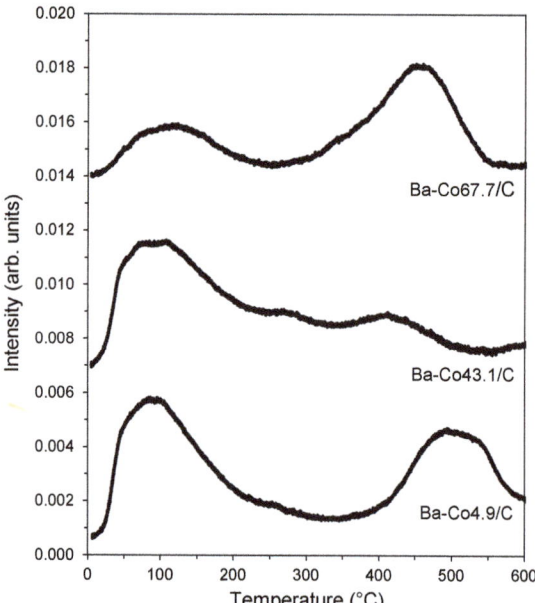

Figure 5. H_2-TPD profiles of the selected catalysts of low (Ba-Co4.9/C), medium (Ba-Co43.1/C), and high (Ba-Co67.7/C) cobalt content.

The results of the quantitative analysis of the obtained desorption profiles are summarized in Table 2. As indicated by the blank experiments, hydrogen was not adsorbed either on the carbon support or the barium promoter. Hence, the adsorbate (H_2) uptake was ascribed entirely to the presence of cobalt. The data revealed that the content of the active phase strongly influenced its adsorption on the support surface and, thus, the metal particle size. It is observed that the more cobalt the catalyst contained, the lower its dispersion (FE) and the greater particle size (d_{Co}). However, using a carbon support of well-developed texture and the proper preparation procedure, a wide range of Co particle sizes was obtained, from 3 nm for the Ba-Co4.9/C catalyst to 45 nm for the Ba-Co67.7/C catalyst. It should also be noted that a high degree of the active phase dispersion on carbon was achieved even with a high metal loading. Comparing the cobalt crystallite size and cobalt particle size (Table 1) with the d_{Co} values determined by the chemisorption method (Table 2), it is noticeable that the d_{Co} values were smaller than those estimated by the XRPD or TEM methods. The difference was especially pronounced for the samples with low cobalt content (4.9 and 9.5 wt% Co). However, it should be remembered that XRPD and TEM measurements were performed ex situ. Due to the tendency of the reduced catalysts to oxidize when contacted with air, the size of Co crystallites and particles estimated by these methods might be overestimated.

Table 2. Chemisorptive characteristics of the Ba-Co/C catalysts.

Catalyst	H_2 Uptake (µmol g_{Co}^{-1})	FE (%) [1]	d_{Co} (nm) [2]
Ba-Co4.9/C	2740	32.5	3
Ba-Co9.5/C	1530	18.0	7
Ba-Co28.2/C	850	9.5	13
Ba-Co43.1/C	540	6.3	20
Ba-Co52.9/C	330	3.9	33
Ba-Co59.4/C	270	3.2	39
Ba-Co67.7/C	240	2.8	45

[1] FE (fraction exposed)—the active phase (cobalt) dispersion defined as the ratio of surface cobalt atoms to total cobalt atoms in the catalyst sample; [2] the average Co particle size in the catalyst sample.

2.3. Catalyst Activity

Figure 6 illustrates the effect of the cobalt particle size (d_{Co}) calculated from hydrogen chemisorption on the activity of the Ba-Co/C catalysts expressed as an average reaction rate (r) and surface reaction rate (TOF). It is clearly visible that the activity of the cobalt catalysts in the NH_3 synthesis reaction depended on the particle size of the active phase. The average reaction rate (r) increased with increasing cobalt particle size (d_{Co}), reached a maximum value for particles of about 20 nm in diameter, and decreased for the bigger particles. In the case of the surface reaction rate (TOF), the highest value was observed for cobalt particles in the range of 20–30 nm. An increase or decrease in the Co particle size beyond this range decreased the TOF value. Moreover, extrapolation of the results towards small particle diameters may suggest that extra fine particles, i.e., smaller than 0.5 nm, might be totally inactive (a critical size).

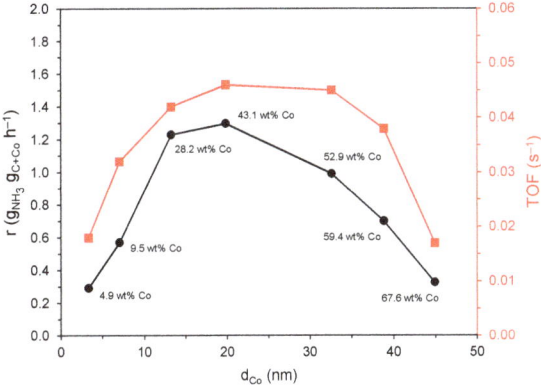

Figure 6. Dependence of activity of the Ba-Co/C catalysts on the cobalt particle size (d_{Co}). Activity expressed as an average NH_3 synthesis reaction rate (r) and surface reaction rate (TOF); measurement conditions: T = 400 °C, p = 9.0 MPa, H_2 + N_2 + NH_3 flow rate 70 dm^3 h^{-1}, H_2/N_2 = 3, x_{NH_3} = 8.5 mol%; TOF determined based on average reaction rate (r) values and the hydrogen chemisorption data. (Values marked by the points indicate the cobalt content in the catalysts corresponding to the crystallites of a given size. Ba content in all the catalysts was 1 mmol g_C^{-1}).

On this basis, it can be concluded that ammonia synthesis conducted on the cobalt catalyst is a structure-sensitive reaction. The obtained dependence (Figure 6) revealed the existence of an optimal size of cobalt particles (20–30 nm), ensuring the highest activity of the catalyst. This observation is of great importance for the implementation of the studied catalysts. It has been found that using a support with a well-developed texture and the proper preparation procedure, the Ba-Co/C cobalt catalysts with high Co content and optimal active phase particle size can be obtained. This, in turn, transfers into a high activity in the ammonia synthesis reaction.

The comparison of the catalytic performance (expressed as the reaction rate) of the selected, most active among the studied catalysts and other Ba-Co/C catalysts previously described in the literature [26,27] is presented in Table 3. Under the same measurement conditions, the studied catalysts showed higher activity than other Ba-Co/C catalysts of similar composition. Moreover, the reaction rate for the studied catalysts is almost four times higher than for the commercial iron catalyst (KM I).

Table 3. The comparison of the reaction rate of ammonia synthesis over the different cobalt catalysts and commercialized iron catalyst (KM I). Reaction conditions: T = 400 °C, p = 9.0 MPa, $H_2 + N_2 + NH_3$ flow rate 70 $dm^3\ h^{-1}$, H_2/N_2 = 3, x_{NH3} = 8.5 mol%.

Catalyst	r_{NH3} ($g_{NH3}\ g_{(C+Co)}^{-1}\ h^{-1}$)	Reference
Ba-Co28.2/C	1.23	This work
Ba-Co43.1/C	1.30	This work
Ba-Co22.8N/B	0.86	[26]
Ba-Co25.4$_{N(R/P+C)}$/B	1.05	[26]
Ba-Co$_{R+P+C}$/C	0.65	[27]
KM I	0.33	[26]

3. Discussion

In trying to explain the correlation between the reactivity of the cobalt surface in ammonia synthesis and the particle size of the active phase, three issues should be considered: (1) the interaction between cobalt particles and carbon support, (2) the structure sensitivity of the ammonia synthesis reaction on cobalt, and (3) the presence of a promoter in the Ba-Co/C catalyst.

(1) Electron-withdrawing functional groups, primarily those containing oxygen, covering the surface of carbon, can interact with cobalt particles causing the decrease of catalytic activity of their surface in the ammonia synthesis reaction. This is due to the fact that electron-poor surfaces are less active in dissociative adsorption of nitrogen, i.e., the rate-determining step of ammonia synthesis. This phenomenon was previously reported by Aika et al. [37–39] for Ru/C systems. The negative impact of oxygen complexes on the surface activity of ruthenium was also observed by Raróg-Pilecka et al. [40] in the reaction of ammonia decomposition over Ru/C catalysts. Presumably, a similar effect may occur for the studied cobalt catalyst supported on carbon. In this respect, it should be expected that the adverse interaction of the oxygen-containing groups would diminish with increasing cobalt particle size and thus result in the increased surface activity of the catalysts expressed as TOF. However, such a relationship was not observed in the discussed systems (Figure 6). A systematic increase in catalyst activity (TOF) was observed for small and medium-size particles ($d_{Co} \leq 20$ nm). In the case of the catalysts with d_{Co} sizes in the range of 20–30 nm, the highest activity was observed, followed by its gradual decrease with a further increase in cobalt particle size. Moreover, the activated carbon used for preparing Ba-Co/C catalysts was subjected to high-temperature processing (1900 °C), which led to a deep purification of carbon, including the removal of surface functional groups. The additional gasification in a mixture of water vapor-argon performed after high-temperature annealing led to the formation of new oxygen groups on the carbon surface. However, as shown in [41], the number of these groups is insignificant. Therefore, the carbon-cobalt interaction cannot explain the TOF vs d_{Co} dependence (Figure 6).

(2) The second possibility of interpreting the correlation between the reactivity of the cobalt surface in ammonia synthesis (TOF) and the Co particle size (d_{Co}) is the structure sensitivity of the ammonia synthesis reaction. As noted in the Section 1, this phenomenon is related to the size of the metal particles on the surface of which the reaction takes place. The results of the conducted research revealed (Figure 6) that there is an optimal size of cobalt particles (20–30 nm), for which the highest activity was obtained in the ammonia synthesis reaction. However, it has been reported in the literature [42] that the maximum concentration of highly coordinated sites (i.e., highly active sites) was obtained for cobalt crystallites with a size of 2–6 nm. The observed discrepancy may result from the assumptions that the metal particles were formed by well-defined crystallites, which shape, morphology and structure did not change with the increase of their diameter. However, the possibility that in the Co/C systems, these structural properties of cobalt crystallites may change as the particles increase cannot be ruled out. According to Kitakami et al. [7], there is a close relationship between the cobalt particle size and the crystal phase. It was shown that particles of diameter ≤20 nm crystallize in the fcc Co phase, whereas there is a mixture of the fcc Co and hcp Co phases in crystallites with a diameter of ca. 30 nm. For particles with a diameter ≥40 nm, the hcp Co phase dominates with only a small amount of the fcc Co phase. According to Rambeau et al. [24], the activity of cobalt in NH_3 synthesis strongly depends on the metal structure and is twice higher for the hcp Co phase than for the fcc Co phase. In this respect, the shift of the optimal Co particle size towards higher values (about 20 nm) seems understandable for the studied catalysts. The hcp Co phase appeared for the particles with a diameter of 20–30 nm, and there was still a significant contribution of the surface cobalt atoms (a sufficient dispersion). However, XRPD studies (Figure 2) revealed that the fcc Co phase dominated in the Ba-Co/C catalysts. The hcp Co phase started to be noticeable in the diffraction patterns of the catalysts with Co particle size (d_{Co}, Table 2) of ca. 20 nm and bigger (the Ba-Co43.1/C and Ba-Co67.7/C, Figure 2). Nevertheless, these studies were performed ex situ. Therefore, the possibility that, under the reaction conditions, a greater amount of the hcp Co phase may be present in the cobalt particles of optimal size cannot be ruled out. Advanced in situ experiments are necessary to clarify this issue thoroughly.

(3) The character of TOF vs. d_{Co} dependence (Figure 6) may presumably in some extent be related to the presence of a promoter in the Ba-Co/C catalyst. The possibility of some characteristic, preferential placement of the barium promoter on cobalt particles of a specific size cannot be ruled out. Assuming that the barium promoter located on the flat surfaces of Co particles causes an increase in the catalytic activity of the adjacent cobalt atoms, a systematic increase of the TOF value with increasing Co particle size should be expected, as one can see in Figure 6. The maximum can be quite broad because it results from the way of particle growth (small particles can be round, but larger particles can be rather flat). A decrease in the activity is observed only for bigger Co particles (≥40 nm), in which the ratio of surface cobalt atoms to all cobalt atoms is very small (low dispersion—Table 2). Thus, the addition of a barium promoter to the Co/C system may result either in a change of activity of active sites already existing on the Co surface or the formation of new sites with particularly advantageous catalytic properties in the ammonia synthesis reaction, but only for the particles of medium size (20–30 nm). The observed, opposite to the expected, phenomenon, i.e., a decrease in activity for cobalt particles of larger size (i.e., particles ≥ 40 nm with a greater fraction of the hcp phase) is an intriguing issue and the explanation is not obvious. In order to confirm the presented hypothesis about the influence of the presence of barium, further and more thorough investigation is required, which may shed light on this issue.

4. Materials and Methods

4.1. Catalyst Preparation

The activated carbon GF40 (Norit, Glasgow, UK) was used as support for preparing the Ba-Co/C catalysts. Briefly, starting carbon (extrusions of 2 mm in diameter) was subjected to two-step modification: (1) heating at 1900 °C for 2 h in argon flow, (2) gasification in water vapor/argon flow at 865 °C for 5 h up to 32 wt% of mass loss. Then, the carbon material was washed with distilled water and dried at 120 °C for 18 h in air.

The Ba-Co/C catalysts of different cobalt content in the range of 4.9 to 67.7 wt%. were synthesized by the wet impregnation method. An appropriate amount of cobalt nitrate hexahydrate ($Co(NO_3)_2 \cdot 6H_2O$, Acros Organics, Thermo Fischer Scientific, Kandel, Germany) was dissolved in ethanol, and carbon support was added to this solution. After impregnation for 24 h, the solvent was evaporated under reduced pressure. Then, calcination in argon flow was carried out at 200 °C for 18 h. For the catalysts containing more than 10 wt% Co, the impregnation procedure and subsequent calcination in argon were repeated until the required Co content was obtained. After the last impregnation, materials were calcined at 200 °C for 18 h in air. To obtain the Ba-promoted catalysts, the calcined samples were impregnated with an aqueous solution of barium nitrate ($Ba(NO_3)_2$, Chempur, Karlsruhe, Germany, 130 g dm^{-3}) at 90 °C for 16 h to obtain a constant and optimal Ba content in all the catalysts equal to 1 mmol g_C^{-1}. Subsequently, the solid materials were separated from the hot solution and dried at 90 °C for 18 h in air. The obtained materials were crushed and sieved to get a 0.2–0.63 mm fraction. The catalyst samples were denoted as Ba-Co(x)/C, where x is the Co content (wt%) in an unpromoted material (estimated based on the mass balance before and after impregnation and final calcination).

4.2. Catalyst Characterization

The textural parameters of carbon support were determined by nitrogen physisorption using an ASAP2020 instrument (Micromeritics Instrument Co., Norcross, GA, USA). Before the measurements, the sample was degassed in vacuum in two stages: at 90 °C for 1 h and then at 300 °C for 4 h. The total surface area and total pore volume were estimated using the Brunauer-Emmett-Teller (BET) and Barrett-Joyner-Halenda (BJH) adsorption isotherm models, respectively.

The phase composition of carbon support and the selected catalysts (the spent catalysts after exposure to air; ex situ measurements) was determined using X-ray powder diffraction (XRPD). Data were collected with a Siemens D5000 diffractometer (München, Germany) in a Bragg–Brentano configuration using a Cu-sealed tube operating at 40 kV and 40 mA with a Ni filter. Each sample was measured in the scattering angle range of 5° to 100° with a 0.02° step and a counting speed of 1.0 s step^{-1}. The average size of cobalt crystallites and graphite/graphite-like packages were determined from Scherrer's equation using the integral width of reflections fitted to the analytical Pearson VII functions.

TEM images were recorded with a Philips CM20 Super Twin microscope (Amsterdam, the Netherlands), which provides a 0.25 nm resolution at 200 kV. The samples for TEM studies were prepared by grinding the materials in a mortar and then dispersing them in methanol. A droplet of the suspension was placed on a microscope grid covered with a perforated carbon film.

The H_2 chemisorption measurements were carried out in a fully automated AutoChem 2920 instrument (Micromeritics Instrument Co., Norcross, GA, USA) equipped with a TCD detector and supplied with high purity (6N) gases. The catalyst sample of 0.2 g was reduced in H_2 flow (40 cm^3 min^{-1}) at 520 °C at a ramping rate of 10 °C min^{-1} for 20 h. Next, the gas flow was switched to argon (40 cm^3 min^{-1}) for purging the sample at 600 °C for 45 min and then cooling the sample to 150 °C. Then, the H_2 adsorption was carried out at 150 °C for 15 min, during cooling the sample to 0 °C and then at 0 °C for 15 min. After rinsing the sample with argon (40 cm^3 min^{-1}) at 0 °C for 1 h to remove weakly adsorbed hydrogen, the catalyst was heated in argon flow (40 cm^3 min^{-1}) to 600 °C with a 10 °C min^{-1} temperature ramp and the concentration of H_2 in the outlet gas was monitored (H_2-TPD). Consequently, the amount of hydrogen evolved to the gas phase was determined by integrating the H_2 desorption profile. The H_2-TPD data were used for calculating the cobalt dispersion expressed as FE (fraction exposed, i.e., the number of surface Co atoms related to the total number of Co atoms) and the average size of cobalt particles

(d_{Co}). The H/Co = 1 stoichiometry [43] and the formula proposed by Borodziński and Bonarowska [44] were used to determine FE and d_{Co}.

4.3. Activity Measurements

The kinetic measurements of NH_3 synthesis were carried out in a differential reactor supplied with a high purity (>99.9999%) stoichiometric H_2-N_2-NH_3 mixture of controlled ammonia concentration (x_1). A detailed description of the apparatus and methodology can be found elsewhere [45,46]. Briefly, under steady-state conditions of temperature (400 °C), gas flow rate (70 dm^3 h^{-1}), pressure (9.0 MPa), and ammonia concentration in the inlet gas (8.5 mol% of NH_3), small increments (x_2) in the concentration of ammonia formed on the catalyst due to the reaction were measured interferometrically. Consequently, the NH_3 synthesis rate was determined. Typically, small catalyst samples (0.5 g, fraction of 0.2–0.63 mm) were used in the studies. Before the experiments, reduction (activation) of the samples was performed in H_2/N_2 = 3 flow (40 dm^3 h^{-1}) at 0.1 MPa following the temperature program: 400 °C (20 h) → 470 °C (24 h) → 520 °C (24 h).

5. Conclusions

A series of barium-promoted cobalt catalysts supported on graphitized carbon was prepared, characterized, and tested in the ammonia synthesis reaction. The active phase particle size was varied from 3 to 45 nm by changing the metal loading in the range of 4.9–67.7 wt%. The characterization of cobalt particles revealed a correlation between the reactivity of the cobalt surface in ammonia synthesis and the particle size of the active phase—the phenomenon of structure sensitivity of the reaction. The dependence of the reaction rate expressed as TOF on the particle size of the active phase indicated that there is an optimal size of cobalt particles (20–30 nm), ensuring the highest activity of the cobalt catalyst in the ammonia synthesis reaction. Increasing or decreasing the particle size caused a decrease in activity, even leading to the expected total loss of catalyst activity for fine Co particles (smaller than 0.5 nm). The size effect may be most likely attributed to changes in Co crystalline structure. For the particles with a diameter of 20–30 nm, the hcp Co phase (more active than the fcc Co phase) appeared, and there was still a significant contribution of the surface cobalt atoms (a sufficient dispersion) conducive to higher activity.

Author Contributions: Conceptualization, A.T., M.Z. and W.R.-P.; methodology, A.T., M.Z. and W.R.-P.; investigation, M.Z., A.T., H.R., W.P., B.M., L.K. and W.R.-P.; writing—original draft preparation, M.Z.; writing—review and editing, M.Z., H.R. and W.R.-P.; visualization, H.R.; supervision, M.Z. and W.R.-P.; funding acquisition, A.T. All authors have read and agreed to the published version of the manuscript.

Funding: This research was financially supported by the National Science Centre, Poland (Research project No. 2016/23/N/ST5/00685).

Data Availability Statement: All data is available within the paper.

Conflicts of Interest: The authors declare no conflict of interest.

References

1. Boudart, M. Catalysis by Supported Metals. *Adv. Catal.* **1969**, *20*, 153–166. [CrossRef]
2. Somorjai, G.A.; Carrazza, J. Structure sensitivity of catalytic reactions. *J. Ind. Eng. Chem. Fundam.* **1986**, *25*, 63–69. [CrossRef]
3. Samorjai, G.A. The structure sensitivity and insensitivity of catalytic reactions in light of the adsorbate induced dynamic restructuring of surfaces. *Catal. Lett.* **1990**, *7*, 169–182. [CrossRef]
4. Van Santen, R.A. Complementary structure sensitive and insensitive catalytic relationships. *Acc. Chem. Res.* **2009**, *42*, 57–66. [CrossRef] [PubMed]
5. Strongin, D.R.; Carrazza, J.; Bare, S.R.; Somorjai, G.A. The importance of C7 sites and surface roughness in the ammonia synthesis reaction over iron. *J. Catal.* **1987**, *103*, 213–215. [CrossRef]
6. Jacobsen, C.J.H.; Dahl, S.; Hansen, P.L.; Törnqvist, E.; Jensen, L.; Topsøe, H.; Prip, D.V.; Møenshaug, P.B.; Chorkendorff, I. Structure sensitivity of supported ruthenium catalysts for ammonia synthesis. *J. Mol. Catal. A Chem.* **2000**, *163*, 19–26. [CrossRef]

7. Kitakami, O.; Sato, H.; Shimada, Y.; Sato, F.; Tanaka, M. Size effect on the crystal phase of cobalt fine particles. *Phys. Rev. B* **1997**, *56*, 13849–13854. [CrossRef]
8. Bozso, F.; Ertl, G.; Grunze, M.; Weiss, M. Interaction of nitrogen with iron surfaces: I. Fe(100) and Fe(111). *J. Catal.* **1977**, *49*, 18–41. [CrossRef]
9. Bozso, F.; Ertl, G.; Weiss, M. Interaction of nitrogen with iron surfaces: II. Fe(110). *J. Catal.* **1977**, *50*, 519–529. [CrossRef]
10. Spencer, N.D.; Schoonmaker, R.C.; Somorjai, G.A. Iron single crystals as ammonia synthesis catalysts: Effect of surface structure on catalyst activity. *J. Catal.* **1982**, *74*, 129–135. [CrossRef]
11. Dumesic, J.A.; Topsøe, H.; Khammouma, S.; Boudart, M. Surface, catalytic and magnetic properties of small iron particles: II. Structure sensitivity of ammonia synthesis. *J. Catal.* **1975**, *37*, 503–512. [CrossRef]
12. Dumesic, J.A.; Topsøe, H.; Boudart, M. Surface, catalytic and magnetic properties of small iron particles: III. Nitrogen induced surface reconstruction. *J. Catal.* **1975**, *37*, 513–522. [CrossRef]
13. Somorjai, G.A.; Materer, N. Surface structures in ammonia synthesis. *Top. Catal.* **1994**, *1*, 215–231. [CrossRef]
14. Zhang, B.-Y.; Su, H.-Y.; Liu, J.-X.; Li, W.-X. Interplay Between Site Activity and Density of BCC Iron for Ammonia Synthesis Based on First-Principles Theory. *ChemCatChem* **2019**, *11*, 1928–1934. [CrossRef]
15. Parker, I.B.; Waugh, K.C.; Bowker, M. On the structure sensitivity of ammonia synthesis on promoted and unpromoted iron. *J. Catal.* **1988**, *114*, 457–459. [CrossRef]
16. Qian, J.; An, Q.; Fortunelli, A.; Nielsen, R.J.; Goddard, W.A., III. Reaction Mechanism and Kinetics for Ammonia Synthesis on the Fe(111) Surface. *J. Am. Chem. Soc.* **2018**, *140*, 6288–6297. [CrossRef]
17. Dahl, S.; Logadottir, A.; Egeberg, R.C.; Larsen, J.H.; Chorkendorff, I.; Törnqvist, E.; Nørskov, J.K. Role of Steps in N_2 Activation on Ru(0001). *Phys. Rev. Lett.* **1999**, *83*, 1814–1817. [CrossRef]
18. Dahl, S.; Törnqvist, E.; Chorkendorff, I. Dissociative adsorption of N_2 on Ru(0001): A surface reaction totally dominated by steps. *J. Catal.* **2000**, *192*, 381–390. [CrossRef]
19. Dahl, S.; Sehested, J.; Jacobsen, C.J.H.; Törnqvist, E.; Chorkendorff, I. Surface science based microkinetic analysis of ammonia synthesis over ruthenium catalysts. *J. Catal.* **2000**, *192*, 391–399. [CrossRef]
20. Van Hardeveld, R.; Van Montfoort, A. The influence of crystallite size on the adsorption of molecular nitrogen on nickel, palladium and platinum: An infrared and electron-microscopic study. *Surf. Sci.* **1966**, *4*, 396–430. [CrossRef]
21. Liang, C.; Wei, Z.; Xin, Q.; Li, C. Ammonia synthesis over Ru/C catalysts with different carbon supports promoted by barium and potassium compounds. *Appl. Catal. A Gen.* **2001**, *208*, 193–201. [CrossRef]
22. Raróg-Pilecka, W.; Miśkiewicz, E.; Szmigiel, D.; Kowalczyk, Z. Structure sensitivity of ammonia synthesis over promoted ruthenium catalysts supported on graphitized carbon. *J. Catal.* **2005**, *231*, 11–19. [CrossRef]
23. Li, L.; Jiang, Y.-F.; Zhang, T.; Cai, H.; Zhou, Y.; Lin, B.; Lin, X.; Zheng, Y.; Zheng, L.; Wang, X.; et al. Size sensitivity of supported Ru catalysts for ammonia synthesis: From nanoparticles to subnanometric clusters and atomic clusters. *Chem* **2022**, *8*, 749–768. [CrossRef]
24. Rambeau, G.; Jorti, A.; Amariglio, H. Catalytic activity of a cobalt powder in NH_3 synthesis in relation with the allotropic transformation of the metal. *J. Catal.* **1985**, *94*, 155–165. [CrossRef]
25. Zhang, B.Y.; Chen, P.-P.; Liu, J.-X.; Su, H.-Y.; Li, W.-X. Influence of Cobalt Crystal Structures on Activation of Nitrogen Molecule: A First-Principles Study. *J. Phys. Chem. C* **2019**, *123*, 10956–10966. [CrossRef]
26. Raróg-Pilecka, W.; Miśkiewicz, E.; Kępiński, L.; Kaszkur, Z.; Kielar, K.; Kowalczyk, Z. Ammonia synthesis over barium-promoted cobalt catalysts supported on graphitized carbon. *J. Catal.* **2007**, *249*, 24–33. [CrossRef]
27. Raróg-Pilecka, W.; Miśkiewicz, E.; Matyszek, M.; Kaszkur, Z.; Kępiński, L.; Kowalczyk, Z. Carbon-supported cobalt catalyst for ammonia synthesis: Effect of preparation procedure. *J. Catal.* **2006**, *237*, 207–210. [CrossRef]
28. Hagen, S.; Barfod, R.; Fehrmann, R.; Jacobsen, C.J.H.; Teunissen, H.T.; Chorkendorff, I. Ammonia synthesis with barium-promoted iron–cobalt alloys supported on carbon. *J. Catal.* **2003**, *214*, 327–335. [CrossRef]
29. Hagen, S.; Barfod, R.; Fehrmann, R.; Jacobsen, C.J.H.; Teunissen, H.T.; Stahl, K.; Chorkendorff, I. New efficient catalyst for ammonia synthesis: Barium-promoted cobalt on carbon. *Chem. Commun.* **2002**, *11*, 1206–1207. [CrossRef]
30. Raróg-Pilecka, W.; Karolewska, M.; Truszkiewicz, E.; Iwanek, E.; Mierzwa, B. Cobalt catalyst doped with cerium and barium obtained by Co-precipitation method for ammonia synthesis process. *Catal. Lett.* **2011**, *141*, 678–684. [CrossRef]
31. Tarka, A.; Patkowski, W.; Zybert, M.; Ronduda, H.; Wieciński, P.; Adamski, P.; Sarnecki, A.; Moszyński, D.; Raróg-Pilecka, W. Synergistic Interaction of Cerium and Barium-New Insight into the Promotion Effect in Cobalt Systems for Ammonia Synthesis. *Catalysts* **2020**, *10*, 658. [CrossRef]
32. Cao, A.; Bukas, V.J.; Shadravan, V.; Wang, Z.; Li, H.; Kibsgaard, J.; Chorkendorff, I.; Nørskov, J.K. A spin promotion effect in catalytic ammonia synthesis. *Nat. Commun.* **2022**, *13*, 2382. [CrossRef]
33. Li, Z.Q.; Lu, C.J.; Xia, Z.P.; Zhou, Y.; Luo, Z. X-ray diffraction patterns of graphite and turbostratic carbon. *Carbon* **2007**, *45*, 1686–1695. [CrossRef]
34. Kowalczyk, Z.; Sentek, J.; Jodzis, S.; Diduszko, R.; Presz, A.; Terzyk, A.; Kucharski, Z.; Suwalski, J. Thermally modified active carbon as a support for catalysts for NH_3 synthesis. *Carbon* **1996**, *34*, 403–409. [CrossRef]
35. Toth, P. Nanostructure quantification of turbostratic carbon by HRTEM image analysis: State of the art, biases, sensitivity and best practices. *Carbon* **2021**, *178*, 688–707. [CrossRef]

36. Ruz, P.; Banerjee, S.; Pandey, M.; Sudarsan, V.; Sastry, P.U.; Kshirsagar, R.J. Structural evolution of turbostratic carbon: Implications in H_2 storage. *Solid State Sci.* **2016**, *62*, 105–111. [CrossRef]
37. Zhong, Z.; Aika, K. The Effect of Hydrogen Treatment of Active Carbon on Ru Catalysts for Ammonia Synthesis. *J. Catal.* **1998**, *173*, 535–539. [CrossRef]
38. Zhong, Z.; Aika, K. Effect of ruthenium precursor on hydrogen-treated active carbon supported ruthenium catalysts for ammonia synthesis. *Inorg. Chem. Acta* **1998**, *280*, 183–188. [CrossRef]
39. Aika, K.; Hori, H.; Ozaki, A. Activation of nitrogen by alkali metal promoted transition metal I. Ammonia synthesis over ruthenium promoted by alkali metal. *J. Catal.* **1972**, *27*, 424–431. [CrossRef]
40. Raróg-Pilecka, W.; Szmigiel, D.; Komornicki, A.; Zieliński, J.; Kowalczyk, Z. Catalytic properties of small ruthenium particles deposited on carbon: Ammonia decomposition studies. *Carbon* **2003**, *41*, 589–591. [CrossRef]
41. Bonarowska, M.; Raróg-Pilecka, W.; Karpiński, Z. The use of active carbon pretreated at 2173 K as a support for palladium catalysts for hydrodechlorination reactions. *Catal. Today* **2011**, *169*, 223–231. [CrossRef]
42. Van Hardeveld, R.; Hartog, F. The statistics of surface atoms and surface sites on metal crystals. *Surf. Sci.* **1969**, *15*, 189–230. [CrossRef]
43. Reuel, R.C.; Bartholomew, C.H. The stoichiometries of H_2 and CO adsorptions on cobalt: Effects of support and preparation. *J. Catal.* **1984**, *85*, 63–77. [CrossRef]
44. Borodziński, A.; Bonarowska, M. Relation between Crystallite Size and Dispersion on Supported Metal Catalysts. *Langmuir* **1997**, *13*, 5613–5620. [CrossRef]
45. Kowalczyk, Z. Effect of potassium on the high pressure kinetics of ammonia synthesis over fused iron catalyst. *Catal. Lett.* **1996**, *37*, 173–179. [CrossRef]
46. Ronduda, H.; Zybert, M.; Patkowski, W.; Tarka, A.; Ostrowski, A.; Raróg-Pilecka, W. Kinetic studies of ammonia synthesis over a barium-promoted cobalt catalyst supported on magnesium–lanthanum mixed oxide. *J. Taiwan Inst. Chem. Eng.* **2020**, *114*, 241–248. [CrossRef]

Article

On Optimal Barium Promoter Content in a Cobalt Catalyst for Ammonia Synthesis

Aleksandra Tarka [1], Magdalena Zybert [1], Hubert Ronduda [1], Wojciech Patkowski [1], Bogusław Mierzwa [2], Leszek Kępiński [3] and Wioletta Raróg-Pilecka [1,*]

[1] Faculty of Chemistry, Warsaw University of Technology, Noakowskiego 3, 00-664 Warsaw, Poland; aleksandra_tarka@interia.pl (A.T.); magdalena.zybert@pw.edu.pl (M.Z.); hubert.ronduda.dokt@pw.edu.pl (H.R.); wpatkowski@ch.pw.edu.pl (W.P.)
[2] Institute of Physical Chemistry, Polish Academy of Sciences, Kasprzaka 44/52, 01-224 Warsaw, Poland; bmierzwa@ichf.edu.pl
[3] Institute of Low Temperature and Structure Research, Polish Academy of Sciences, Okólna 2, 50-950 Wrocław, Poland; l.kepinski@intibs.pl
* Correspondence: wiola@ch.pw.edu.pl; Tel.: +48-22-2345766

Abstract: High priority in developing an efficient cobalt catalyst for ammonia synthesis involves optimizing its composition in terms of the content of promoters. In this work, a series of cobalt catalysts doubly promoted with cerium and barium was prepared and tested in ammonia synthesis (H_2/N_2 = 3, 6.3 MPa, 400 °C). Barium content was studied in the range of 0–2.6 mmol g_{Co}^{-1}. Detailed characterization studies by nitrogen physisorption, SEM-EDX, XRPD, H_2-TPR, and H_2-TPD showed the impact of barium loading in CoCeBa catalysts on the physicochemical properties and activity of the catalysts. The most pronounced effect was observed in the development of the active phase surface, a differentiation of weakly and strongly binding sites on the catalyst surface and changes in cobalt surface activity (TOF). Barium content in the range of 1.1–1.6 mmol g_{Co}^{-1} leads to obtaining a catalyst with the most favorable properties. Its excellent catalytic performance is ascribed to the appropriate Ba/Ce molar ratio, i.e., greater than unity, which results in not only a structural promotion of barium, but also a modifying action associated with the in-situ formation of the $BaCeO_3$ phase.

Keywords: ammonia synthesis; cobalt catalyst; barium; promoter; optimization

1. Introduction

Many industrial processes require the use of catalysts to carry out a reaction at a suitable rate and under desirable conditions. A classic example of heterogeneous catalysis is ammonia synthesis over Fe- or Ru-based catalysts. These metals alone are almost inactive in ammonia synthesis [1–3], but their activity significantly increases in the presence of some compounds. These compounds, added to catalysts in small amounts, are called promoters, and they play a crucial role in heterogeneous catalysis [4]. They improve catalyst properties by enhancing activity, lifespan (long-term stability), and selectivity. Promoters can be divided into structural and electronic promoters, depending on the mode of their action. Structural promoters primarily increase the catalyst's activity by increasing the surface area of an active phase. Electronic (chemical) promoters increase the catalytic activity by modifying the active metal and by increasing the reaction rate per surface area [5,6]. This is a general description, but the function of each promoter is always specific to the particular catalytic system and the particular reaction.

In the case of a fused iron catalyst for ammonia synthesis, aluminum oxide, calcium oxide, and magnesium oxide are typically used as structural promoters [7]. They stabilize the active planes of the metal (role of Al_2O_3), increase and stabilize the catalyst surface area during reduction (role of CaO and MgO), and increase the catalyst resistance to impurities (role of CaO). Moreover, potassium oxide is used as an electronic promoter. It

can increase the rate-limiting step of dissociative nitrogen adsorption [8] or decrease the concentration of produced ammonia adsorbed on the iron surface, and hence make more active sites available for nitrogen [9]. In the case of ruthenium, alkali metals are electronic promoters whose influence is similar to that noted for the iron catalyst [2,10–12]. High activity in ammonia synthesis was also achieved by promoting ruthenium by cesium and barium [13,14].

Among the alkaline earth metals, barium is of particular attention as a very effective promoter of catalysts for the synthesis of ammonia [15–27]. Its role is significant and has been thoroughly investigated by many researcher groups, but its effect has not been fully explained. Some authors have shown that it is a structural promoter [18–20], whereas others postulate that it exhibits an electronic effect [21–23]. There is also a viewpoint in which the influence of barium may have a mixed character, i.e., both structural and electronic [15,24,27]. A cobalt catalyst doubly promoted with cerium and barium was the subject of our previous research [15,27]. These cobalt–cerium–barium systems exhibited very high activity in ammonia synthesis. The studies revealed that the double promotion of cobalt with Ce and Ba causes an approximately twofold increase in catalyst activity, compared to the cobalt system promoted only with barium, and over tenfold increase in activity compared to the cobalt system doped only with cerium. The particularly beneficial properties of the catalyst result from the synergistic action of the two promoters. Cerium oxide is a structural promoter in cobalt–cerium–barium systems preventing Co particles from sintering during the reaction and stabilizing the active hcp cobalt phase [15,27–29]. Optimal cerium oxide content (1.0 mmol g_{Co}^{-1}), i.e., one which provides the most favorable catalytic properties, was determined during our further studies [28]. In the case of barium, although it mainly exhibits an electronic character, structural effects have been observed. However, the most important is the participation in the in-situ formation (under the conditions of catalysts activation) of the $BaCeO_3$ phase. It is the third promoter with strong electron-donating properties and the ability to differentiate the structure of hydrogen adsorption sites (co-existence of weakly and strongly binding sites) on the active phase surface. However, these observations were carried out only for one catalyst composition (Ce content 1.0 mmol g_{Co}^{-1}, Ba content 1.4 mmol g_{Co}^{-1}) [15,27].

As a continuation of the systematic studies of barium-promoted cobalt catalysts, in this work, we studied ammonia synthesis on doubly promoted cobalt–cerium–barium catalysts of various barium content (in the range of 0–2.6 mmol g_{Co}^{-1}). The main goal was to determine the optimal content of the barium promoter, providing the most favorable catalytic properties of the studied CoCeBa systems. Thorough characterization studies of the prepared materials by nitrogen physisorption, Scanning Electron Microscopy with Energy Dispersive Spectroscopy (SEM-EDX), X-ray Powder Diffraction (XRPD), Temperature-Programmed Reduction with hydrogen (H_2-TPR), and Temperature-Programmed Desorption of hydrogen (H_2-TPD) were used to determine the influence of the barium content on the properties and catalytic performance of the doubly promoted cobalt catalysts in ammonia synthesis.

2. Results and Discussion

2.1. Textural Characteristics (N_2 Physisorption)

The textural characteristics of the catalyst precursors are summarized in Table 1. A small addition of the barium promoter (0.2 mmol g_{Co}^{-1}) results in a decrease of the specific surface area (S_{BET}) of the precursor by about 11% and an over twofold decrease of the total pore volume (V_P) (CoCeBa(0.2)), compared to that of the precursor without barium (CoCe). When the barium content in samples is increased to 1.4 mmol Ba g_{Co}^{-1}, a further decrease in S_{BET} and V_P values is observed, which is probably a result of filling pores with the barium salt. In the samples containing 1.6 mmol Ba g_{Co}^{-1} and more, changes in textural parameters (S_{BET}, V_P) are negligibly small. Selected precursors of small, medium, and high Ba content were reduced in-situ, and their specific surface areas were measured (Table 1, S_R values). A significant decrease in a specific surface area of the materials is observed due to reduction. For example, the surface of the CoCe sample decreases after reduction over 11 times, and in the case of CoCeBa(2.6), the specific surface area after reduction is nearly 22-times smaller than before the reduction. For CoCeBa(1.4)e, the specific surface area after reduction was only 5 times lower. This indicates that barium has a beneficial effect when added in an optimal amount and effectively prevents sintering of the grains during reduction. The increase of the specific surface area with an increase of the barium content is observed for samples containing 0.2–1.4 mmol Ba g_{Co}^{-1}. The S_R value for the reduced sample promoted by a small amount of barium (CoCeBa(0.2)) is approximately 9% larger than the surface area of the reduced sample without barium (CoCe). The highest S_R value after reduction is observed for CoCeBa(1.4). Further increase of barium content, i.e., over 1.4 mmol Ba g_{Co}^{-1}, caused a decrease in the surface area of the reduced samples. The observed effects indicate that barium may behave as a structural promoter. However, there is an optimum content of Ba, which may develop the catalyst surface. After exceeding it, the catalyst grains sinter, resulting in decrease of the specific surface area of the catalysts.

Table 1. Chemical composition and textural parameters of the promoted cobalt catalysts.

Catalyst	Ba Content [1] (mmol g_{Co}^{-1})	Ba/Ce Molar Ratio [2]	S_{BET} [3] ($m^2\,g^{-1}$)	S_R [4] ($m^2\,g^{-1}$)	V_P [5] ($cm^3\,g^{-1}$)
CoCe	0.00	-	85	7.5	0.34
CoCeBa(0.2)	0.20	0.2	76	8.2	0.15
CoCeBa(0.5)	0.48	0.4	67	-	0.14
CoCeBa(1.1)	1.05	0.9	63	-	0.14
CoCeBa(1.4)	1.36	1.2	52	10.7	0.12
CoCeBa(1.6)	1.61	1.4	52	-	0.12
CoCeBa(2.0)	1.95	1.7	53	5.0	0.15
CoCeBa(2.2)	2.19	2.0	52	4.9	0.15
CoCeBa(2.6)	2.62	2.3	52	2.4	0.14

[1] Values determined based on mass balance after impregnation of the $Co_3O_4 + CeO_2$ sample. [2] Cerium content is constant and equal to 1.1 mmol g_{Co}^{-1}, the value calculated based on the cerium oxide content in the $Co_3O_4 + CeO_2$ sample determined using TG-MS. [3] S_{BET}–specific surface area estimated based on the BET isotherm model. [4] S_R–specific surface area estimated based on the BET isotherm model after hydrogen activation. [5] V_P–total pore volume estimated based on the BJH isotherm model.

2.2. Reduction Behavior of the Studied Catalysts (H_2-TPR)

In order to investigate the effect of the barium promoter content on the reducibility of the cobalt catalysts, temperature-programmed reduction measurements were performed. Figure 1 shows the reduction profiles for the studied catalysts. The area of the graph presented in Figure 1 was divided into areas marked as I, II, IIIa, and IIIb to simplify the description of the obtained signals. In the reduction profile of the CoCe sample, which in the oxidized form is a mixture of Co_3O_4 and CeO_2 oxides, two peaks (marked in Figure 1 as II and III) are observed, with maxima at 289 °C and 479 °C, respectively. These signals correspond to a two-step reduction of cobalt oxide to metallic cobalt [30,31], in accordance with Equations (1) and (2):

$$Co_3O_4 + H_2 \rightarrow 3CoO + H_2O \quad (1)$$

$$3CoO + 3H_2 \rightarrow 3Co + 3H_2O \quad (2)$$

Under the measurement conditions (temperature increase from 30 to 700 °C at a constant rate of 10 °C min^{-1}, 10 vol.% H_2/Ar), cerium (IV) oxide did not undergo reduction, which has been reported in previous studies [15,28]. It is noted that the introduction of the barium promoter to the systems containing cobalt (II,III) oxide and cerium (IV) oxide causes a change in the course of their reduction (Figure 1). The TPR profiles of the samples containing barium show a small peak (I) of constant area and maximum in the range of 210 °C ± 15 °C, which may be related to the decomposition of the barium salt. The position of peak II is not influenced by the barium content in the system. Its maximum occurs at the temperature of 299 °C ± 8 °C. However, in the case of the samples containing barium in the amount of 1.4 mmol g_{Co}^{-1} and more, the intensity of peak II slightly increases. Presumably, it results from the overlapping of the peaks related to the reduction of Co_3O_4 to CoO and further decomposition of the barium salt. In the profiles of the samples containing 0.2–1.1 mmol Ba g_{Co}^{-1}, the maximum of peak III shifts towards higher temperatures with increasing barium promoter content. For the samples containing 1.4 mmol Ba g_{Co}^{-1} and more, two peaks (IIIa and IIIb) are observed instead of one in the region where peak III is present. The maximum of peak IIIa occurs in a constant temperature range, i.e., 435 °C ± 5 °C, while the maximum of peak IIIb shifts from the position at 554 °C for CoCeBa (1.4) towards lower temperatures with increasing barium content. Moreover, with the addition of more barium promoter, the decrease in the intensity of peak IIIb is observed, accompanied by the increase in the intensity of peak IIIa. Finally, in the reduction profile of CoCeBa(2.6), a very high intensity of peak IIIa and a negligibly small IIIb peak are observed.

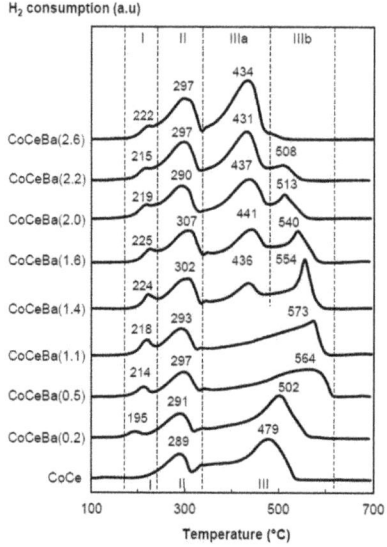

Figure 1. Reduction profiles of the cobalt catalyst promoted with cerium and with a barium loading in the range: 0–2.6 mmol g_{Co}^{-1} (measurement conditions: 30–700 °C, 10° C min^{-1}, 10 vol.% H_2/Ar).

It can be clearly stated that the addition of barium in the range of 0.2–2.2 mmol Ba g_{Co}^{-1} hinders the process of the CoCeBa catalysts reduction. It requires a longer time and ends at a higher temperature than in the case of a sample without barium (CoCe). Additional peaks and complexity of the CoCeBa catalyst precursor reduction profiles are most likely related to the decomposition of barium salts.

2.3. Chemisorption Characteristics of the Active Phase Surface (H_2-TPD)

The profiles of hydrogen desorbing from the surface of the promoted cobalt catalysts with different barium content are presented in Figure 2. The hydrogen desorption curves obtained for the samples containing 0–1.1 mmol Ba g_{Co}^{-1} consist of one broad peak in the low-temperature range (α), extending from about 50 °C to 550 °C. Its maximum is observed at a temperature of about 170 °C. In the profiles of the samples containing 1.4 mmol Ba g_{Co}^{-1} and more (i.e., in the cases where barium is in molar excess to cerium), apart from the low-temperature peak (α), a high-temperature peak (β) appears with a maximum in the range 520–550 °C. The low-temperature peak shifts slightly towards lower temperatures as the barium content in the samples increases. In the case of the high-temperature peak, the maximum changes its position slightly. However, there is no clear trend of this change. The low-temperature signal (α) corresponds to the desorption of hydrogen weakly bound to the cobalt surface. In contrast, the high-temperature signal (β) corresponds to the desorption of H_2 strongly interacting with cobalt [32]. This means that samples containing barium in the content range of 0–1.1 mmol Ba g_{Co}^{-1} have only weak hydrogen-binding sites on their surface. In contrast, on the surface of cobalt in the samples containing 1.4 mmol Ba g_{Co}^{-1} and more, both weakly and strongly hydrogen-binding sites coexist. As the barium content increases, the intensity of the low-temperature peak decreases. The area ratio (β/α, Table 2) of peaks corresponding to strongly and weakly-binding sites on the surface of catalysts increases with barium content above 1.4 mmol Ba g_{Co}^{-1}, and reaches the highest value for the CoCeBa(2.2) system. Moreover, for the two systems with the highest barium content, CoCeBa(2.2) and CoCeBa(2.6), the high-temperature (β) peak begins to dominate the low-temperature (α) one in terms of the area. The observed phenomena, i.e., peak sharpening, slight temperature shifts of their location, appearance of new peaks, indicate the restructuration of the cobalt systems surface, occurring with the increase in barium content. Not only does the number of hydrogen-binding sites change, but the homogenization of their energy and formation of new types of sites also becomes visible. Therefore, it may be stated that these results support the conclusion that barium exhibits the role of a structural promoter in the studied cobalt catalysts.

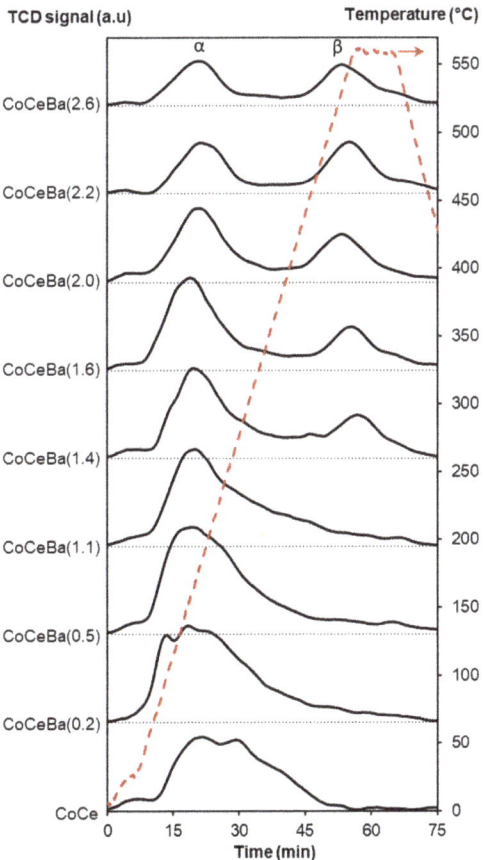

Figure 2. Profiles of hydrogen desorption from the surface of the cobalt catalyst promoted with cerium and with a barium loading in the range: 0–2.6 mmol g_{Co}^{-1}.

Table 2. Chemisorption characteristics of the promoted cobalt catalysts.

Catalyst	H$_2$ Uptake (µmol g$_{Co}^{-1}$)		β/α Peak Area Ratio	S$_{Co}$ [1] (m^2 g$_{Co}^{-1}$)	d$_{Co\text{-}TPD}$ [2] (nm)
	α Peak	β Peak			
CoCe	121.4	-	-	7.7	88
CoCeBa(0.2)	162.5	-	-	10.3	66
CoCeBa(0.5)	160.8	-	-	10.2	66
CoCeBa(1.1)	154.4	-	-	9.8	69
CoCeBa(1.4)	100.1	61.4	0.6	10.2	66
CoCeBa(1.6)	102.7	59.4	0.6	10.2	66
CoCeBa(2.0)	83.2	62.9	0.8	9.2	73
CoCeBa(2.2)	46.5	65.0	1.4	7.1	96
CoCeBa(2.6)	43.6	48.0	1.1	5.6	117

[1] S$_{Co}$—surface area of the active phase (cobalt) estimated based on H$_2$-TPD measurement results.
[2] d$_{Co\text{-}TPD}$—average cobalt particle size estimated based on H$_2$-TPD measurement results.

Based on the hydrogen desorption curves (Figure 2) and calculated H$_2$ uptake (Table 2), the average size of metallic cobalt particles (d$_{Co\text{-}TPD}$) and the active phase surface area (S$_{Co}$) were determined. The data presented in Table 2 show that even a small addition of the barium promoter (0.2 mmol Ba g$_{Co}^{-1}$) results in a significant, i.e., about 33%, increase in the active phase surface, compared to that of the catalyst without the barium promoter (sample CoCe). The addition of a larger amount of barium (in the range of 0.5–1.6 mmol Ba g$_{Co}^{-1}$) does not significantly change the Co surface area—for all these systems, the S$_{Co}$ value is constant and amounts to approx. 10 m^2 g$_{Co}^{-1}$. However, a further increase in the content of the barium promoter, i.e., above 1.6 mmol Ba g$_{Co}^{-1}$, causes the surface of the active phase to gradually decrease. The largest cobalt particle size (d$_{Co\text{-}TPD}$) was determined, and thus the lowest metallic cobalt surface (S$_{Co}$) was observed for CoCeBa(2.6), which has the highest Ba content.

2.4. Phase Composition of the Precursors and Catalysts in the Reduced form (XRPD)

The phase composition of the selected promoted cobalt catalysts was analyzed using XRPD. The materials in the oxidized (catalyst precursors) and reduced (catalysts) forms were investigated. The recorded diffraction patterns are presented in Figure 3. Reflexes from Co$_3$O$_4$ are visible in the diffraction patterns of all oxidized samples (Figure 3a). However, there are no reflexes from CeO$_2$. This may indicate the presence of a weakly crystallized or amorphous and/or highly dispersed cerium oxide. The presence of signals from two different Ba-containing phases is observed. Barium nitrate signals are clearly visible for the samples with high barium content. Barium nitrate was used for impregnation and the introduction of a substantial amount of this salt could cause a crystallization of this compound in the form of larger particles, detectable by XRPD. For the samples with low barium content, the amount of salt could be too small to form particles of a size appropriate for XRPD analysis or they were better dispersed within the samples. For the catalysts in the reduced form (Figure 3b), signals derived from Ba(NO$_3$)$_2$ are not detected. According to the literature reports [33], Ba(NO$_3$)$_2$ is transformed into amorphous BaO$_x$ species under ammonia synthesis reaction conditions. The subsequent reaction of BaO$_x$ species with atmospheric CO$_2$ could cause the formation of BaCO$_3$ particles (ex-situ XRPD measurements for the catalyst samples removed from the ammonia synthesis reactor). Drying barium nitrate at 120 °C should not cause the decomposition of the compound. According to the author of [34], the decomposition of pure barium nitrate occurs above 530 °C. Nevertheless, the dispersion of barium nitrate on the surface of another material significantly lowers the decomposition temperature. During drying of the catalyst precursors, the dispersed salt could presumably be partially decomposed into BaO, and due to contact with air (containing CO$_2$), it could transform into BaCO$_3$. Hence, there are visible signals of this phase in the catalyst precursor samples (Figure 3a).

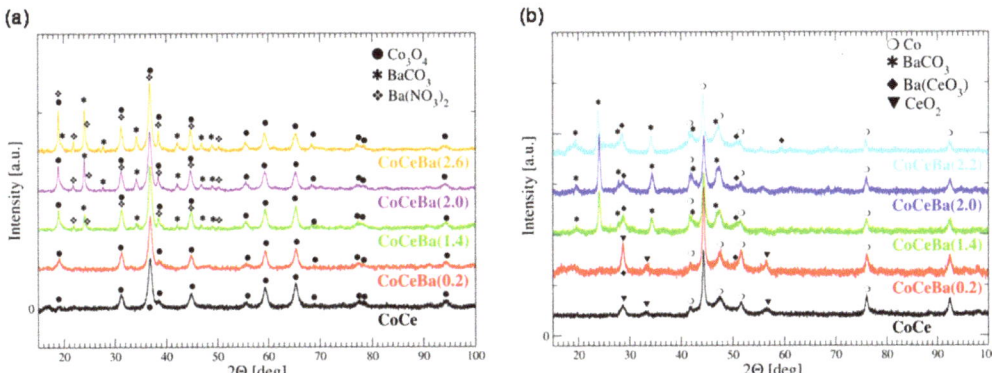

Figure 3. XRPD patterns of the cobalt catalysts promoted with cerium (CoCe) or cerium and barium (CoCeBa) in the form of a precursor (**a**) and the reduced form (**b**).

In the case of reduced samples, metallic cobalt is observed (Figure 3b). There are visible signals typical for hexagonal close-packed cobalt (Co hcp: 41.6° and 47.5°) and face-centered cubic cobalt (Co fcc, 51.5°). The reflexes at the 2θ angles = 44.4°, 75.9°, and 92.4° may come from both phases—Co hcp and Co fcc. It is also worth noting that the form of promoter results from the interaction between barium and cerium compounds in the samples. In CoCeBa(0.2), the cerium promoter is present in the form of two chemical compounds: cerium oxide (CeO_2) and barium cerate ($BaCeO_3$), whereas barium is only observed in the form of $BaCeO_3$. As indicated in Table 1, in sample CoCeBa(0.2), cerium is present in molar excess to barium, so the phase composition determined by XRPD is consistent with the chemical composition. For other samples (i.e., CoCeBa(1.4), CoCe(2.0), and CoCeBa(2.2)), where the Ba/Ce molar ratio is greater than unity (Table 1), no cerium oxide phase is observed, as Ce is likely to be bound entirely in the form of barium cerate.

Based on the results obtained from XRPD measurements, the average cobalt oxide crystallite size ($d_{Co3O4-XRD}$) for the precursor samples and the average metal cobalt crystallite size (d_{Co-XRD}) for the reduced samples (ex-situ measurements) were estimated and are presented in Table 3. The average size of cobalt oxide crystallites ($d_{Co3O4-XRD}$) in all the samples is similar and equals approx. 11 nm. This result may suggest that barium has no structure-forming effect on cobalt in the case of oxidized materials, i.e., it does not increase or decrease the surface area of the cobalt oxide. The estimated average crystallite size of the metallic cobalt in the reduced samples (d_{Co-XRD}) are also similar (in the range of 21–26 nm) and much lower than the d_{Co-TPD} values calculated based on the chemisorption measurements (Table 2). This may be because the XRPD method can determine small cobalt crystallites, structurally ordered fragments of larger aggregates (agglomerates). However, during chemisorption measurements, only the outer surface of the particles is available for the adsorbate. Consequently, the values of d_{Co-TPD} related to cobalt particles may be greater than the values of d_{Co-XRD} related to cobalt crystallites.

2.5. Morphology and Element Distribution of the Catalysts in the Reduced form (SEM-EDX)

Figure 4 contains SEM images of the selected catalysts in the reduced form. They show that the morphology of CoCe and CoCeBa(0.2) samples is similar. Both materials consist of nanoparticles formed into larger grains. Although the images show the surface morphology regardless of its composition, they confirm previous observations and conclusions drawn for the active phase of the catalyst from H_2-TPD and XRPD analyses (Table 2—d_{Co-TPD} values and Table 3—d_{Co-XRD} values, respectively). It was then found that the differences observed between the cobalt crystallite sizes estimated based on these two methods result from the fact that the crystallites of the active phase with an ordered crystal structure (detectable by the XRPD method) may aggregate into larger particles (agglomerates). Their size resulting from the development of their external surface, accessible to gaseous probe molecules, is estimated based on H_2-TPD data. The SEM analysis also confirms that CoCeBa(1.4) is a catalyst with a well-developed surface. In fact, it has the most developed surface among the samples tested with this method, which is in good agreement with the results of textural studies of the reduced form of this sample. It can also be seen (Figure 4) that CoCeBa(2.2) differs in morphology from the other samples due to high barium content. The particles which form the grains of the catalyst containing 2.2 mmol Ba g_{Co}^{-1} are much larger than in the case of the other tested catalysts. Consequently, the sample's surface is smaller, i.e., less developed. These observations are also consistent with the results of the specific surface area after reduction (Table 1—S_R values) and active phase surface area (Table 2, S_{Co} values) obtained for the studied catalysts.

Table 3. The crystallite sizes of Co-containing phases of the promoted cobalt catalysts.

Catalyst	$d_{Co_3O_4\text{-XRD}}$ [1] (nm)	$d_{Co\text{-XRD}}$ [2] (nm)
CoCe	10	22
CoCeBa(0.2)	11	24
CoCeBa(1.4)	12	21
CoCeBa(2.0)	11	21
CoCeBa(2.2)	-	26
CoCeBa(2.6)	10	-

[1] The mean cobalt oxide crystallite size ($d_{Co_3O_4\text{-XRD}}$) for the precursor samples. [2] The mean metal cobalt crystallite size ($d_{Co\text{-XRD}}$) for the reduced samples (ex-situ measurements).

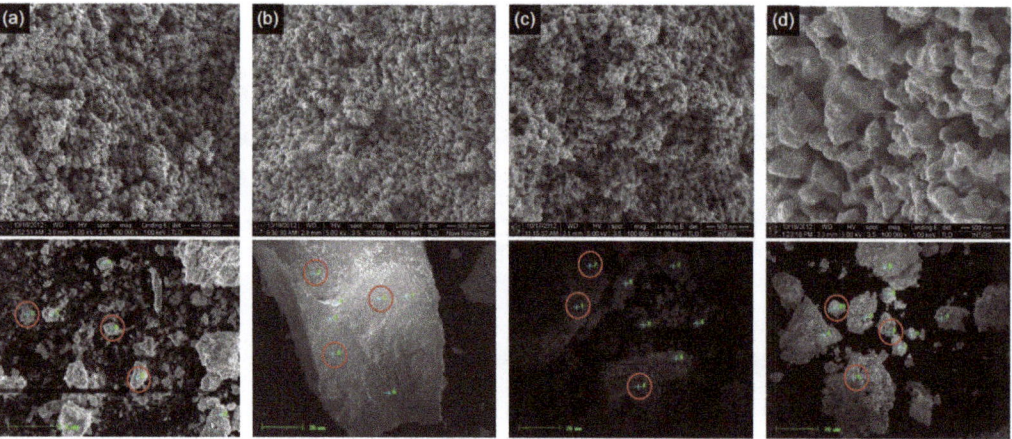

Figure 4. SEM images and corresponding BSE images of selected promoted cobalt catalysts in the reduced form (ex-situ measurement): (**a**) CoCe, (**b**) CoCeBa(0.2), (**c**) CoCeBa(1.4), (**d**) CoCeBa(2.2). The selected grains taken into account in the element distribution analysis (EDX) are marked in red.

The distribution of Co, Ce, and Ba elements on the surface of the tested catalyst samples in their reduced form was determined using EDX analysis. The results of the relative ratios of the elements in three randomly selected points (Figure 4) on the surface of the samples are presented in Table 4. In all the tested catalysts, at each of the selected measuring points, the Co/Ce ratio is similar, which indicates a uniform distribution of cobalt and cerium on the catalyst surface. This is due to the preparation of the catalyst precursor by co-precipitation, which ensures a good distribution of the cerium promoter throughout the sample. However, in the Ba-promoted samples, the distribution of barium with respect to cobalt is not uniform. The Co/Ba ratio discrepancies between selected points may be due to the method of sample preparation, i.e., incipient wetness impregnation method, which does not ensure uniform deposition of the promoting element on the catalyst surface.

Table 4. Distribution of Co, Ce, and Ba (relative ratio of the elements in three randomly selected points) on the surface of the promoted cobalt catalysts.

Catalyst	Point	Element Share (%)			Elements Ratio	
		Co	Ce	Ba	Co/Ce	Co/Ba
CoCe	1.	88.5	11.5	-	7.7	-
	2.	88.4	11.6	-	7.6	-
	3.	88.0	12.0	-	7.3	-
CoCeBa(0.2)	1.	85.7	11.3	3.0	7.6	28.6
	2.	86.8	11.5	1.7	7.5	51.1
	3.	85.6	11.3	3.1	7.6	27.6
CoCeBa(1.4)	1.	79.4	9.0	11.7	8.8	6.8
	2.	77.9	9.1	13.0	8.6	6.0
	3.	69.2	7.8	23.0	8.9	3.0
CoCeBa(2.2)	1.	79.1	9.9	11.1	8.0	7.1
	2.	75.3	8.8	15.9	8.6	4.7
	3.	51.5	6.3	2.2	8.2	23.4

2.6. Activity in NH$_3$ Synthesis (Catalytic Activity Measurements)

Measurements of the catalyst activity were carried out, and the average reaction rate (r_{av}) of ammonia synthesis was determined. Based on r_{av} values and H$_2$ uptake values (from H$_2$-TPD measurements), the surface activity of the catalyst, expressed as the turnover frequency (TOF), was determined. The results are shown in Figure 5.

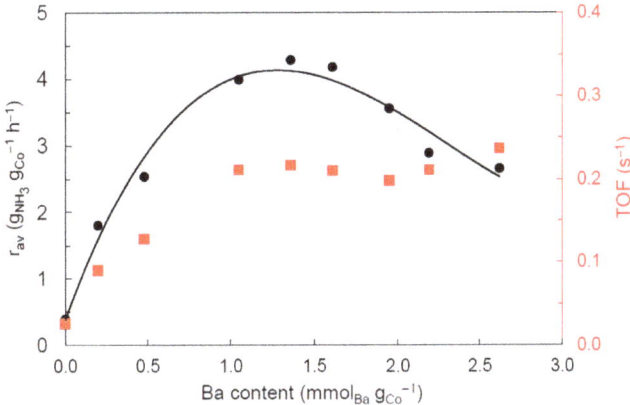

Figure 5. Dependence of activity of the promoted cobalt catalysts on the barium promoter content. Activity expressed as an average NH$_3$ synthesis reaction rate (r_{av}, ●) and TOF (■); measurement conditions: T = 400 °C, p = 6.3 MPa, H$_2$/N$_2$ = 3; TOF was determined based on r_{av} values and the hydrogen chemisorption data.

The addition of a small amount of barium (0.2 mmol Ba g_{Co}^{-1}) results in an almost five-fold increase in average reaction rate and a 3.5-fold increase in the TOF value, compared to the catalysts without barium (CoCe). With a further increase in barium content, i.e., in the range of 0.2–1.4 mmol Ba g_{Co}^{-1}, the r_{av} value increases, which may be a direct result of the development of the cobalt surface (S_{Co}, Table 2) in the catalysts. The average reaction rate (r_{av}) reaches a maximum value for CoCeBa(1.4), and decreases with a further increasing of the barium content. The gradual increase in the activity with the addition of the barium promoter is a result of an electronic effect of barium. Its presence causes a donation of electrons to the cobalt surface, which then facilitates the cleavage of adsorbed dinitrogen. This function of the alkaline dopant was also indicated in our previous studies of the discussed cobalt systems [15,27,35], other cobalt catalysts [17,36,37], and ruthenium catalysts for ammonia synthesis [2,11,12,19–24]. Moreover, it is worth nothing that no sign of deactivation of the studied catalysts was observed after overheating (600 °C, 72 h), indicating that all the catalysts display stable performance—a critical parameter, especially in industrial processes. When analyzing the surface activity of the catalysts, it should be noted that the TOF value initially increases with increasing barium content in the samples. Then, for samples containing barium in the range of 1.1–2.6 mmol Ba g_{Co}^{-1}, it reaches a constant value of approx. $0.2\ s^{-1}$. Thus, the decrease in the average reaction rate observed for the samples with the highest barium content may be related to the substantial decrease of the active phase surface (S_{Co}, Table 2).

The superior performance of the CoCeBa(1.4) catalyst for ammonia synthesis is revealed by comparison to the literature results with similar reaction conditions (Table 5). It can be seen that the ruthenium catalysts display higher NH_3 synthesis rates, compared to the iron and cobalt catalysts. However, the activity of the CoCeBa(1.4) catalyst is much higher than that of the commercial fused iron catalyst (about three-fold). Thus, it might be considered as a valuable alternative to the iron catalyst for ammonia synthesis.

Table 5. The comparison of NH_3 synthesis rate (r_{NH3}) over cobalt, iron, and ruthenium catalysts under the pressure of about 6 MPa and 400 °C.

Catalyst	r_{NH3} ($g_{NH3}\ g_{cat}^{-1}\ h^{-1}$)	Reference
Fe	1.2	[11]
Co@BaO/MgO-700red	3.1	[38]
Ru/CeO$_2$	5.4	[39]
K-Ru/C	4.4	[11]
CoCeBa(1.4)	3.4	This work

Summarizing the presented results, it should be stated that the catalytic properties of the cobalt systems doubly promoted with cerium and barium strictly depend on the content of barium. When the cerium promoter is present in molar excess to barium (Ba/Ce < 1, Table 1), barium acts as a typical structural promoter. It prevents the sintering of cobalt particles during reduction, causing the development of the active phase surface and thus an increase in the activity of the catalysts in ammonia synthesis. However, in cases where the barium to cerium ratio is greater than unity (Ba/Ce > 1, Table 1), the modifying (electronic) character of the barium promoter is also revealed. It was observed that despite the decrease in cobalt surface, the surface activity (TOF values) of the catalysts containing more than 1.4 mmol Ba g_{Co}^{-1} remained at a high and stable level. However, considering our previous investigation of the synergistic effect of the cerium and barium promoters in the cobalt catalyst [15,28], the properties of the cobalt–cerium–barium systems should also be related to the presence of the BaCeO$_3$ phase. For barium-rich catalysts (Ba/Ce > 1), the binding of the entire cerium promoter in the form of BaCeO$_3$ ensures that in all these systems, the amount of this third promoter, which exhibits a strong electron-donating effect on the cobalt surface, is similar. This is reflected by the nearly constant TOF value, indicating a similar surface activity of the active phase of these materials. The lack of a free cerium promoter in the form of CeO$_2$ (i.e., not bound in BaCeO$_3$) causes the decay of the structural influence of cerium, which explains the decrease in the active phase surface area (Table 2). The effect of the decrease in the cobalt surface for the barium-rich catalysts may also be associated with the phenomenon of surface enrichment with barium, in which the barium promoter introduced in excess in relation to cerium may accumulate on the cobalt particles, blocking the access of the reagents to the active sites of the catalyst. This phenomenon was previously observed in our studies of cobalt systems promoted with barium [36]. Based on the results of the conducted experiments, the optimal barium promoter content in the CoCeBa catalyst was established. The most favorable properties were obtained for the catalytic systems containing 1.1–1.6 mmol Ba g_{Co}^{-1}.

3. Materials and Methods

3.1. Preparation of the Catalysts

In the first step, a mixture of cobalt and cerium oxides (Co_3O_4 + CeO_2) was prepared using the co-precipitation method and subsequent calcination. Appropriate amounts of cobalt(II) nitrate hexahydrate and cerium(III) nitrate hexahydrate were dissolved in distilled water. Excess potassium carbonate aqueous solution was slowly added under continuous stirring to the nitrates solution until the pH was 9. Both of the solutions were first heated to 90 °C. The obtained precipitate was filtered and washed with cold distilled water until the pH was neutral. It was then dried at 120 °C in air for 18 h and calcined at 500 °C in air for 18 h. Afterwards, the material was impregnated with various amounts of barium using an aqueous solution of barium(II) nitrate (incipient wetness impregnation) and dried in air at 120 °C. Finally, the samples were crushed and sieved to obtain grain size in the range of 0.20–0.63 mm. The last step of catalyst preparation was the reduction of precursors carried out directly before measurements, which required a reduced form of the materials and before the catalytic activity studies (details can be found below in a characterization methods description, Section 3.2.). As a result, a series of doubly promoted cobalt catalysts were obtained of cerium content equal to 1.1 mmol g_{Co}^{-1}, while the barium content varied in the range of 0–2.6 mmol g_{Co}^{-1}. Cerium content was determined using thermal analysis coupled with mass spectrometry according to the methodology described in [40] for the precursor containing only Co_3O_4 and CeO_2 (i.e., the precursor before impregnation with barium). The basis of the discussed method is the fact that under the measurement conditions (heating in argon), cerium oxide is stable, whereas cobalt (II,III) oxide decomposes to cobalt (II) oxide at a temperature of about 750 °C. The recorded mass loss allows the determination of the Co_3O_4 content in a mixed oxide system. The rest of the sample consists of CeO_2. The content of barium promoter in the final precursor samples (before reduction) was calculated based on the mass balance before and after impregnation with barium salt of the precursor samples containing Co_3O_4 and CeO_2. Barium content and a molar ratio of barium to cerium are listed in Table 1. Materials are denoted as CoCeBa(n), where n is the amount of barium in relation to cobalt, as indicated in Table 1. The sample without barium, donated as CoCe, was a reference material.

3.2. Catalyst Characterisation

The specific surface area of the precursors (i.e., materials in the unreduced form), total pore volume, and specific surface area of the reduced form of the selected materials were determined by nitrogen physisorption with an ASAP2020 instrument (Micromeritics Instrument Co., Norcross, GA, USA). Before the measurements, each sample of the precursors was degassed in vacuum in two stages: at 90 °C for 1 h and then at 200 °C for 4 h. Before the measurement for the selected materials in their reduced form, the precursors were reduced in-situ at 550 °C for 10 h in hydrogen flow and then subjected to degassing at 150 °C for 2 h. The reduction and degassing were conducted in the apparatus directly before N_2 physisorption measurements.

The morphology and element distribution for the selected catalytic materials in the reduced form was studied using scanning electron microscopy (SEM) coupled with energy-dispersive X-ray spectroscopy EDX (FEI NovaNanoSEM 230, FEI Company, Hillsboro, OR, USA).

The phase composition of the selected precursors and the catalysts in the reduced form were determined using X-ray powder diffraction (XRPD). Data were collected with a Rigaku-Denki Geigerflex (Rigaku Denki Co., Ltd., Tokyo, Japan) diffractometer in Brag–Brentano configuration using CuKα radiation. The samples were scanned in a 2θ range of 15–100° with a step of 0.02° and counting time 5 s. The average size of Co_3O_4 crystallites (in the precursors) and metallic cobalt crystallites (in the reduced catalysts) was estimated based on the Scherrer equation using the integral width of the reflex filled to the analytical Pearson VII function.

A reducibility of the catalyst precursors was studied using Temperature-Programmed Reduction with hydrogen (H_2-TPR) at AutoChem2920 (Micromeritics Instrument Co.). Samples of the precursors containing about 0.03 g of Co_3O_4 were heated from room temperature to 700 °C at a constant rate of 10 °C min^{-1} in the flow of 10 vol.% H_2/Ar (40 mL min^{-1}). The hydrogen consumption was measured by a Thermal Conductivity Detector (TCD).

The catalysts' active phase surface was characterized using temperature-programmed hydrogen desorption (H_2-TPD) using a PEAK-4 instrument. Measurements were conducted in a flow set-up supplied with high purity (99.99995 vol.%) gases (total gas flow rate 40 mL min^{-1}) in a quartz U-tube reactor. Samples of the catalyst precursors containing 0.5 g Co_3O_4 + CeO_2 were reduced in a flow of H_2/Ar = 4:1 mixture (40 mL min^{-1}) at 550 °C for 18 h. The system was then flushed with flowing argon at 570 °C for 1 h and cooled to 150 °C. The H_2 adsorption was carried out at 150 °C for 15 min, then continued during cooling of the sample to 0 °C and for 15 min at 0 °C. After flushing with Ar to remove weakly bound molecules of H_2, the temperature was increased to 550 °C at a constant rate (10 °C min^{-1}) and then kept for 10 min at 550 °C while monitoring the concentration of hydrogen desorbing from the surface of the catalyst. The surface area of the active phase (S_{Co}) and average cobalt particle size (d_{Co}) were calculated assuming H/Co = 1 stoichiometry of hydrogen adsorption [41].

3.3. Catalytic Tests

The activity of the catalysts in ammonia synthesis was tested in a tubular flow reactor under steady-state conditions (6.3 MPa, 400 °C, H_2/N_2 = 3, gas flow rate 70 dm^3 h^{-1}). Before the activity measurements, samples of the catalyst precursors (grain size 0.2–0.63 mm) of about 0.5 g were activated in a high purity H_2/N_2 = 3 mixture (99.99995 vol%., gas flow rate 30 dm^3 h^{-1}) under atmospheric pressure in accordance with the temperature program: 470 °C for 72 h, then 520 °C for 24 h and finally 550 °C for 48 h. The product concentration in the outlet gas was measured interferometrically. The catalytic activity was determined and expressed as an average NH_3 synthesis reaction rate (r_{av}). A detailed description of the set-up and the method for calculating the reaction rate was described in [42]. Moreover, the activity of the catalyst surface expressed as TOF was estimated. The calculation was based on the values of the average reaction rates (r_{av}) and the number of active sites on the cobalt surface determined during chemisorption measurements (H_2-TPD).

4. Conclusions

In summary, the influence of barium content on the physicochemical properties and catalytic activity of the cobalt catalyst doubly promoted with cerium and barium was investigated. A series of catalysts of various barium promoter content in the range of 0–2.6 mmol g_{Co}^{-1} was prepared, characterized, and tested in ammonia synthesis. The dual nature of the role of the barium promoter(structural and modifying) was revealed, but it strictly depends on the barium-to-cerium molar ratio. For systems of the Ba/Ce molar ratio lower than unity (Ba/Ce < 1), the structural character of barium was observed. It manifested itself mainly in preventing sintering of the active phase during reduction. For the best catalytic performance of the CoCeBa system, the Ba/Ce molar ratio should be greater than unity (Ba/Ce > 1), which results in not only a structural promotion of barium, but also a modifying action associated with the in-situ formation of the $BaCeO_3$ phase. It was primarily reflected in the differentiation of weakly and strongly binding sites on the catalyst surface and changes of the cobalt surface activity (TOF). The optimal barium content in the range of 1.1–1.6 mmol g_{Co}^{-1} leads to obtaining a catalyst with the most favorable properties. Its excellent catalytic performance is ascribed to the appropriate Ba/Ce molar ratio. It is also related to the presence of the $BaCeO_3$ phase, which plays the role of a third promoter of a high electron-donating character.

Author Contributions: Conceptualization, A.T., M.Z. and W.R.-P.; methodology, A.T., M.Z. and W.R.-P.; investigation, A.T., M.Z., H.R., W.P., B.M., L.K. and W.R.-P.; writing—original draft preparation, A.T. and M.Z.; writing—review and editing, M.Z., H.R. and W.R.-P.; visualization, A.T., and H.R.; supervision, M.Z.; funding acquisition, W.R.-P. All authors have read and agreed to the published version of the manuscript.

Funding: The research has been funded by The National Centre for Research and Development within The Applied Research Programme, grant No. PBS2/A1/13/2014.

Data Availability Statement: All data is available within the paper.

Acknowledgments: The authors thank Ewa Iwanek from the Faculty of Chemistry, Warsaw University of Technology, for additional proofreading and language corrections.

Conflicts of Interest: The authors declare no conflict of interest.

References

1. Tamaru:, K. The history of the development of ammonia synthesis. In *Catalytic Ammonia Synthesis: Fundamentals and Practice*; Jennings, J.R., Ed.; Plenum Press: New York, NY, USA, 1991; pp. 1–18.
2. Aika, K.; Hori, H.; Ozaki, A. Activation of nitrogen by alkali metal promoted transition metal I: Ammonia synthesis over ruthenium promoted by alkali metal. *J. Catal.* **1972**, *27*, 424–431. [CrossRef]
3. Kowalczyk, A.; Sentek, J.; Jodzis, S.; Mizera, E.; Góralski, J.; Paryjczak, T.; Diduszko, R. An alkali-promoted ruthenium catalyst for the synthesis of ammonia, supported on thermally modified active carbon. *Catal. Lett.* **1997**, *45*, 65–72. [CrossRef]
4. Hutchings, G.J. Promotion in Heterogeneous Catalysis: A Topic Requiring a New Approach? *Catal. Lett.* **2001**, *75*, 1–12. [CrossRef]
5. Stoltz, P. Structure and surface chemistry of industrial ammonia synthesis catalysts. In *Ammonia. Catalysis and Manufacture*; Nielsen, A., Ed.; Springer-Verlag: Berlin, Germany, 1995; pp. 17–102.
6. Vayenas, C.G.; Bebelis, S.; Pliangos, C.; Brosda, S.; Tsiplakides, D. Promotion in Heterogeneous Catalysis. In *Electrochemical Activation of Catalysis*; Kluwer Academic Publishers: New York, NY, USA, 2002; pp. 15–90.
7. Rase, H.F. Ammonia Converter. In *Handbook of Commercial Catalysts: Heterogeneous Catalysts*; CRC Press: Boca Raton, FL, USA, 2000; pp. 449–454.
8. Ertl, G.; Lee, S.B.; Weiss, M. Adsorption of nitrogen on potassium promoted Fe(111) and (100) surfaces. *Surf. Sci.* **1982**, *114*, 527–545. [CrossRef]
9. Strongin, D.R.; Somorjai, G.A. The effects of potassium on ammonia synthesis over iron single-crystal surfaces. *J. Catal.* **1988**, *109*, 51–60. [CrossRef]
10. Aika, K.; Kubota, J.; Kadowaki, Y.; Niwa, Y.; Izumi, Y. Molecular sensing techniques for the characterization and design of new ammonia catalysts. *Appl. Surf. Sci.* **1997**, *121–122*, 488–491. [CrossRef]
11. Raróg, W.; Kowalczyk, Z.; Sentek, J.; Składanowski, D.; Zieliński, J. Effect of K, Cs and Ba on the kinetics of NH_3 synthesis over carbon-based ruthenium catalysts. *Catal. Lett.* **2000**, *68*, 163–168. [CrossRef]
12. Rossetti, I.; Pernicone, N.; Forni, L. Promoters effect in Ru/C ammonia synthesis catalyst. *Appl. Catal. A* **2001**, *208*, 271–278. [CrossRef]
13. Forni, L.; Molinari, D.; Rossetti, I.; Pernicone, N. Carbon-supported promoted Ru catalyst for ammonia synthesis. *Appl. Catal. A* **1999**, *185*, 269–275. [CrossRef]
14. Kowalczyk, Z.; Krukowski, M.; Raróg-Pilecka, W.; Szmigiel, D.; Zielinski, J. Carbon-based ruthenium catalyst for ammonia synthesis: Role of the barium and caesium promoters and carbon support. *Appl. Catal. A* **2003**, *248*, 67–73. [CrossRef]
15. Raróg-Pilecka, W.; Karolewska, M.; Truszkiewicz, E.; Iwanek, E.; Mierzwa, B. Cobalt catalyst doped with cerium and barium obtained by co-precipitation method for ammonia synthesis process. *Catal. Lett.* **2011**, *141*, 678–684. [CrossRef]
16. Hagen, S.; Barfod, R.; Fehrmann, R.; Jacobsen, C.J.H.; Teunissen, H.T.; Chorkendorff, I. Ammonia synthesis with barium-promoted iron–cobalt alloys supported on carbon. *J. Catal.* **2003**, *214*, 327–335. [CrossRef]
17. Tarka, A.; Zybert, M.; Truszkiewicz, E.; Mierzwa, B.; Kępiński, L.; Moszyński, D.; Raróg-Pilecka, W. Effect of a Barium Promoter on the Stability and Activity of Carbon-Supported Cobalt Catalysts for Ammonia Synthesis. *ChemCatChem* **2015**, *7*, 2836–2839. [CrossRef]
18. Zhong, Z.; Aika, K. The Effect of Hydrogen Treatment of Active Carbon on Ru Catalysts for Ammonia Synthesis. *J. Catal.* **1998**, *173*, 535–539. [CrossRef]
19. Bielawa, H.; Hinrichsen, O.; Birkner, A.; Muhler, M. The Ammonia-Synthesis Catalyst of the Next Generation: Barium-Promoted Oxide-Supported Ruthenium. *Angew. Chem. Int. Ed.* **2001**, *40*, 1061–1063. [CrossRef]
20. Szmigiel, D.; Bielawa, H.; Kurtz, M.; Hinrichsen, O.; Muhler, M.; Raróg, W.; Jodzis, S.; Kowalczyk, Z.; Znak, L.; Zielinski, J. The Kinetics of Ammonia Synthesis over Ruthenium-Based Catalysts: The Role of Barium and Cesium. *J. Catal.* **2002**, *205*, 205–212. [CrossRef]
21. Hansen, T.W.; Wagner, J.B.; Hansen, P.L.; Dahl, S.; Topsøe, H.; Jacobsen, C.J.H. Atomic-Resolution in Situ Transmission Electron Microscopy of a Promoter of a Heterogeneous Catalyst. *Science* **2001**, *294*, 1508–1510. [CrossRef]

22. Hansen, T.W.; Hansen, P.L.; Dahl, S.; Jacobsen, C.J.H. Support Effect and Active Sites on Promoted Ruthenium Catalysts for Ammonia Synthesis. *Catal. Lett.* **2002**, *84*, 7–12. [CrossRef]
23. Zeng, H.S.; Inazu, K.; Aika, K. The Working State of the Barium Promoter in Ammonia Synthesis over an Active-Carbon-Supported Ruthenium Catalyst Using Barium Nitrate as the Promoter Precursor. *J. Catal.* **2002**, *211*, 33–41. [CrossRef]
24. Truszkiewicz, E.; Raróg-Pilecka, W.; Szmidt-Szałowski, K.; Jodzis, S.; Wilczkowska, E.; Łomot, D.; Kaszkur, Z.; Karpiński, Z.; Kowalczyk, Z. Barium-promoted Ru/carbon catalyst for ammonia synthesis: State of the system when operating. *J. Catal.* **2009**, *265*, 181–190. [CrossRef]
25. Ronduda, H.; Zybert, M.; Patkowski, W.; Ostrowski, A.; Jodłowski, P.; Szymański, D.; Kępiński, L.; Raróg-Pilecka, W. A high performance barium-promoted cobalt catalyst supported on magnesium–lanthanum mixed oxide for ammonia synthesis. *RSC Adv.* **2021**, *11*, 14218–14228. [CrossRef]
26. Ronduda, H.; Zybert, M.; Patkowski, W.; Tarka, A.; Ostrowski; Raróg-Pilecka, W. Kinetic studies of ammonia synthesis over a barium-promoted cobalt catalyst supported on magnesium–lanthanum mixed oxide. *J. Taiwan Inst. Chem. Eng.* **2020**, *114*, 241–248. [CrossRef]
27. Tarka, A.; Patkowski, W.; Zybert, M.; Ronduda, H.; Wieciński, P.; Adamski, P.; Sarnecki, A.; Moszyński, D.; Raróg-Pilecka, W. Synergistic Interaction of Cerium and Barium-New Insight into the Promotion Effect in Cobalt Systems for Ammonia Synthesis. *Catalysts* **2020**, *10*, 658. [CrossRef]
28. Karolewska, M.; Truszkiewicz, E.; Mierzwa, B.; Kępiński, L.; Raróg-Pilecka, W. Ammonia synthesis over cobalt catalysts doped with cerium and barium. Effect of the ceria loading. *Appl. Catal. A* **2012**, *445–446*, 280–286. [CrossRef]
29. Lin, S.S.Y.; Kim, D.H.; Ha, S.Y. Metallic phases of cobalt-based catalysts in ethanol steam reforming: The effect of cerium oxide. *Appl. Catal. A* **2009**, *355*, 69–77. [CrossRef]
30. Lin, H.Y.; Chen, Y.W. The mechanism of reduction of cobalt by hydrogen. *Mater. Chem. Phys.* **2004**, *85*, 171–175. [CrossRef]
31. Xue, L.; Zhang, C.; He, H.; Teraoka, Y. Catalytic decomposition of N_2O over CeO_2 promoted Co_3O_4 spinel catalyst. *Appl. Catal. B* **2007**, *75*, 167–174. [CrossRef]
32. Ronduda, H.; Zybert, M.; Patkowski, W.; Ostrowski, A.; Jodłowski, P.; Szymański, D.; Kępiński, L.; Raróg-Pilecka, W. Development of cobalt catalyst supported on $MgO–Ln_2O_3$ (Ln=La, Nd, Eu) mixed oxide systems for ammonia synthesis. *Int. J. Hydrog. Energy* **2022**, *47*, 6666–6678. [CrossRef]
33. Nishi, M.; Chen, S.Y.; Takagi, H. Mild ammonia synthesis over Ba-promoted Ru/MPC catalysts: Effect of the Ba/Ru ratio and the mesoporous structure. *Catalysts* **2019**, *9*, 480. [CrossRef]
34. Bardwell, C.J.; Bickley, R.I.; Poulston, S.; Twigg, M.V. Thermal decomposition of bulk and supported barium nitrate. *Thermochim. Acta* **2015**, *613*, 94–99. [CrossRef]
35. Patkowski, W.; Kowalik, P.; Antoniak-Jurak, K.; Zybert, M.; Ronduda, H.; Mierzwa, B.; Próchniak, W.; Raróg-Pilecka, W. On the effect of flash calcination method on the characteristics of cobalt catalysts for ammonia synthesis process. *Eur. J. Inorg. Chem.* **2021**, *15*, 1518–1529. [CrossRef]
36. Zybert, M.; Wyszyńska, M.; Tarka, A.; Patkowski, W.; Ronduda, H.; Mierzwa, B.; Kępiński, L.; Sarnecki, A.; Moszyński, D.; Raróg-Pilecka, W. Surface enrichment phenomenon in the Ba-doped cobalt catalyst for ammonia synthesis. *Vacuum* **2019**, *168*, 108831. [CrossRef]
37. Tarka, A.; Zybert, M.; Kindler, Z.; Szmurło, J.; Mierzwa, B.; Raróg-Pilecka, W. Effect of precipitating agent on the properties of cobalt catalysts promoted with cerium and barium for NH_3 synthesis obtained by co-precipitation. *Appl. Catal. A* **2017**, *532*, 19–25. [CrossRef]
38. Sato, K.; Miyahara, S.; Tsujimaru, K.; Wada, Y.; Toriyama, T.; Yamamoto, T.; Matsumura, S.; Inazu, K.; Mohri, H.; Iwasa, T.; et al. Barium Oxide Encapsulating Cobalt Nanoparticles Supported on Magnesium Oxide: Active Non-Noble Metal Catalysts for Ammonia Synthesis under Mild Reaction Conditions. *ACS Catal.* **2021**, *11*, 13050–13061. [CrossRef]
39. Fang, B.; Liu, F.; Zhang, C.; Li, C.; Ni, J.; Wang, X.; Lin, J.; Lin, B.; Jiang, L. Sacrificial Sucrose Strategy Achieved Enhancement of Ammonia Synthesis Activity over a Ceria-Supported Ru Catalyst. *ACS Sustain. Chem. Eng.* **2021**, *9*, 8962–8969. [CrossRef]
40. Zybert, M.; Truszkiewicz, E.; Mierzwa, B.; Raróg-Pilecka, W. Thermal analysis coupled with mass spectrometry as a tool to determine the cobalt content in cobalt catalyst precursors obtained by co-precipitation. *Thermochim. Acta* **2014**, *584*, 31–35. [CrossRef]
41. Reule, R.C.; Bartholomew, C.H. The stoichiometries of H_2 and CO adsorptions on cobalt: Effects of support and preparation. *J. Catal.* **1984**, *85*, 63–77. [CrossRef]
42. Kowalczyk, Z. Effect of potassium on the high-pressure kinetics of ammonia synthesis over fused iron catalysts. *Catal Lett.* **1996**, *37*, 173–179. [CrossRef]

Article

Thermal Stability of Potassium-Promoted Cobalt Molybdenum Nitride Catalysts for Ammonia Synthesis

Paweł Adamski *, Wojciech Czerwonko and Dariusz Moszyński

Department of Inorganic Chemical Technology and Environment Engineering, Faculty of Chemical Technology and Engineering, West Pomeranian University of Technology in Szczecin, Piastów Ave. 42, 71-065 Szczecin, Poland; wojciech.czerwonko@zut.edu.pl (W.C.); dmoszynski@zut.edu.pl or dariusz.moszynski@zut.edu.pl (D.M.)
* Correspondence: adamski_pawel@zut.edu.pl; Tel.: +48-91-449-4024

Abstract: The application of cobalt molybdenum nitrides as ammonia synthesis catalysts requires further development of the optimal promoter system, which enhances not only the activity but also the stability of the catalysts. To do so, elucidating the influence of the addition of alkali metals on the structural properties of the catalysts is essential. In this study, potassium-promoted cobalt molybdenum nitrides were synthesized by impregnation of the precursor $CoMoO_4 \cdot 3/4H_2O$ with aqueous KNO_3 solution followed by ammonolysis. The catalysts were characterized with the use of XRD and BET methods, under two conditions: as obtained and after the thermal stability test. The catalytic activity in the synthesis of ammonia was examined at 450 °C, under 10 MPa. The thermal stability test was carried out by heating at 650 °C in the same apparatus. As a result of ammonolysis, mixtures of two phases: Co_3Mo_3N and Co_2Mo_3N were obtained. The phase concentrations were affected by potassium admixture. The catalytical activity increased for the most active catalyst by approximately 50% compared to non-promoted cobalt molybdenum nitrides. The thermal stability test resulted in a loss of activity, on average, of 30%. Deactivation was caused by the collapse of the porous structure, which is attributed to the conversion of the Co_2Mo_3N phase to the Co_3Mo_3N phase.

Keywords: cobalt molybdenum nitrides; ammonia synthesis; phase composition; specific surface area

1. Introduction

The Haber–Bosch process developed in the early years of the twentieth century had a great influence on the production of ammonia. This process uses an iron catalyst that allows direct bonding of H_2 and N_2 and can be considered efficient; however, due to the huge worldwide production of ammonia, the continuous improvement of catalysts is required for both financial and environmental reasons [1]. An important approach to solving this problem was the application of ruthenium-based catalysts. They exhibit high activity in ammonia synthesis but are burdened by several technological flaws, as well as high cost [2,3]. These led researchers to focus on less expensive and more available materials instead of noble metal-based catalysts [4]. Transition metal nitrides proved to be very effective catalysts that can be obtained as binary [5], ternary [1] and very recently quaternary [6] systems that—with further improvements—can serve as excellent catalysts for ammonia synthesis.

Ternary transition metal nitrides are well-known catalysts for hydrogenation reactions such as hydrosulfurization of thioorganic compounds [7–9], hydroprocessing of organic compounds [10] hydrazine decomposition [11] and NO reduction [12]. For ammonia synthesis, theoretical studies indicate that cobalt molybdenum nitride is the most active among the other chemical substances [13,14].

The studies of cobalt molybdenum nitride promoted by alkali metals were carried out by Kojima and Aika [15]. During these studies, a substantial increase in the catalytic activity, caused by the presence of cesium or potassium in the catalyst, was observed. A specific

concentration of each of the promoters was required to obtain the optimal catalytic activity. The cesium promoter was found to be more effective compared to potassium [16]. The proposed beneficiary effect of alkali metals on catalyst activity was described in the terms of changes in the electronic properties of the active sites through the electron-donation mechanism [17]. However, studies of other catalytical systems suggest that the effect of the addition of alkali metals is more complicated.

A structural effect of promoter addition was observed in cobalt molybdenum nitride catalysts. In our previous studies [18,19] we reported the complexity of the phase composition of the cobalt molybdenum nitride catalysts, especially the occurrence of two active phases: Co_2Mo_3N and Co_3Mo_3N. Their concentration ratio varies with the type and concentration of the applied promoters. Cobalt molybdenum nitride catalysts promoted with chromium [19,20] as well as potassium and chromium [21] were compared. The chromium admixture was found to hinder, and potassium to facilitate the formation of the Co_2Mo_3N phase. The mechanism that leads to the coexistence of the Co_2Mo_3N and Co_3Mo_3N phases was studied in detail elsewhere [22]. The Co_2Mo_3N phase was shown to be an intermediate phase in the formation of the Co_3Mo_3N phase. Catalysts with higher concentrations of the Co_2Mo_3N phase are catalytically more active [18,19].

In addition to the catalytical activity, an important factor for the practical application of catalysts is their thermal stability. Each catalyst changes in the course of the reaction, and its deactivation occurs. For an industrial application, the interval between loads of fresh catalyst should be as long as possible, for example, for the iron catalyst, it is about ten years [23]. In the case of cobalt molybdenum nitrides, the available information about stability during prolonged test runs is limited. The stability of the catalytic properties was examined in detail for the non-promoted and cesium-promoted catalysts by Kojima and Aika [24]. It was shown that during the 24 h process under the atmosphere of the reaction gas $N_2 + 3H_2$ at 600 °C, both the unpromoted catalyst and the Cs-promoted catalysts reached the maximum of their activity after 12 h. Subsequently, their activity gradually decreased. At the same time, their BET surface area decreased by approximately half. The increase in activity was associated with the transformation of the intermediate Co and Mo_2N phases into the Co_3Mo_3N phase.

The stability of the unpromoted cobalt molybdenum nitride catalyst was examined via in situ XRD studies [25]. The phase concentrations of Co_2Mo_3N and Co_3Mo_3N under a reaction gas atmosphere at 700 °C were stable. The Co_3Mo_3N phase decomposed into the Co_6Mo_6N phase under a hydrogen atmosphere. This transformation was explained in terms of the high reactivity of bulk lattice nitrogen present in η-carbide structured Co_3Mo_3N by Daisley et al. [26].

The outlook on the behavior under industrial reaction conditions of non-promoted and K, Cs, Cr-promoted catalysts before and after thermostability test was given by Nadziejko et al. [27]. It was shown that the specific surface area and activity of the Cs-promoted catalysts after the thermostability test decreased the most among the studied catalysts. A serious deactivation for Cs-promoted catalysts was also observed by us, in a detailed study [28]. It was associated with the sintering of the crystallites of the active phases and decreasing of the surface area of the catalysts. Furthermore, in the study of Boisen et al. [5] the influence between the activity of the Cs-promoted catalyst and its surface area was observed. Despite their initial high activity, Cs-promoted catalysts are regarded as highly prone to sintering and are inefficient in the long run.

There are still ambiguities regarding the influence of alkali metals on ammonia synthesis catalysts. In addition to electron transfer from the alkali to the active center, other factors such as the change of surface structure or change of the crystallite sizes must be considered. To address this problem in the present study, the thermal stability of the K-promoted catalyst was examined in detail. Cobalt molybdenum nitrides promoted by potassium were studied as catalysts in ammonia synthesis at 450 °C and subsequently after prolonged heating under a H_2-N_2 gas mixture at 650 °C. The transformation of the phase structure, porous structure, and catalytic activity of the examined catalysts were related to the concentration of potassium in the catalytic material.

2. Results

Cobalt molybdenum nitride catalysts modified with an admixture of potassium (henceforth abbreviated as COMON catalysts), were obtained in a three-step process. First, the precipitation of cobalt molybdates, referred to as precursors, was carried out. The precipitation of precursors was followed by their impregnation with the potassium nitrate. Subsequently, the activation process of the catalyst precursors was performed under the flow of pure ammonia (a process called ammonolysis). The obtained catalysts were studied in two chemical states: after ammonolysis of the precursors, and after the activity measurements followed by the aging of the catalysts under increased temperature. More detailed experimental conditions are described in the Section 4.

The X-ray diffraction pattern of the synthesized precursor is consistent with previously published results [20], where the precursor was identified as $CoMoO_4 \cdot 3/4H_2O$ (PDF 04-011-8282). In Figure 1, the XRD pattern acquired after ammonolysis is shown for the exemplary sample containing 0.2 wt.% of potassium. The diffraction reflections observed for the catalysts after ammonolysis, as well as after thermal stability tests, were ascribed exclusively to Co_3Mo_3N or Co_2Mo_3N phases. No oxidic phases, metallic cobalt or Mo_2N were detected, although the surface of all samples after each process was passivated in diluted oxygen prior to XRD analysis.

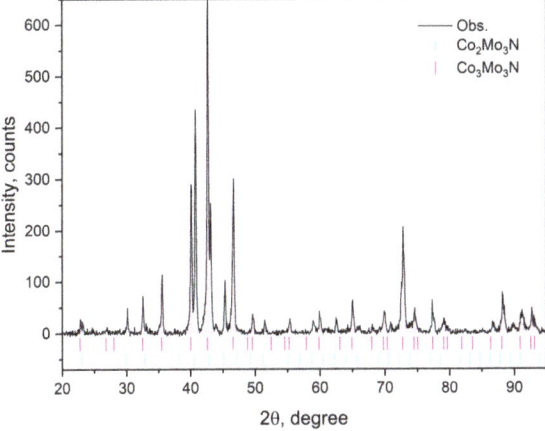

Figure 1. X-ray diffraction pattern of the COMON catalyst containing 0.2 wt.% of potassium acquired after ammonolysis, with the indicated Rietveld refinement.

The weight fractions of the Co_2Mo_3N (PDF 04-010-6426) and Co_3Mo_3N (PDF 04-008-1301) phases identified in the catalysts were determined by X-ray diffraction analysis with the use of Rietveld refinement. The weight fraction of the Co_2Mo_3N phase related to the potassium concentration is shown in Figure 2. The Co_3Mo_3N phase complements the composition of the catalysts. The non-promoted sample contains about 18 wt.% of the cobalt-lean Co_2Mo_3N phase. The content of this phase in the catalyst grows to about 47 wt.% with increasing potassium concentration to top at 0.8 wt.%. At even higher potassium concentrations, a decrease in Co_2Mo_3N concentration is observed. At a potassium concentration of 3.5 wt.%, the Co_2Mo_3N concentration dropped to 18 wt.% of the catalyst.

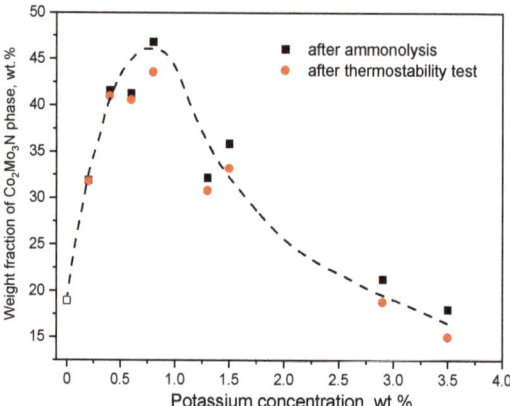

Figure 2. Weight fraction of the Co$_2$Mo$_3$N phase as a function of potassium concentration in COMON catalysts. A line is given for eye guidance.

The catalysts after the thermal stability tests were also subjected to a diffraction analysis, as presented in Figure 2. In general, the relation between the weight fraction of the Co$_2$Mo$_3$N phase and the potassium concentration is similar to that observed for these materials before aging. For the non-promoted catalyst, the share of the Co$_2$Mo$_3$N phase after the aging process was unchanged. However, for the promoted catalysts, the phase composition of the samples changed slightly after the thermostability test and the weight fraction of the Co$_2$Mo$_3$N phase was reduced for all promoted catalysts.

The specific surface area is an important factor that influences the catalytic properties of the catalysts. During the present study, this parameter was measured twice for all of the catalysts: after ammonolysis and after thermostability test. The results are depicted in Figure 3. The specific surface area of fresh, non-promoted COMON catalyst is 15.5 m^2/g, a value typical for many carrier-free metallic catalysts [29,30]. The surface area of the remaining catalysts depends on the concentration of potassium. It amounted to 9.2 m^2/g for the fresh catalyst containing 0.2 wt.% of potassium and increased in the range of potassium concentration between 0.4 wt.% and 0.8 wt.%. The maximum was observed at 0.8 wt.% of potassium and amounted to 10.7 m^2/g. A further increase of potassium concentration resulted in a decrease in the specific surface area. The most prominent loss was observed for catalysts containing 2.9 wt.% and 3.5 wt.% of potassium, where the specific surface area measured was approximately 5.5 m^2/g.

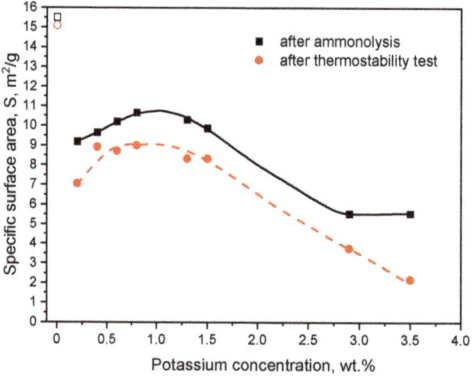

Figure 3. Specific surface area of COMON catalysts as a function of potassium concentration, measured after ammonolysis and after all catalytic tests. Lines are given for eye guidance.

After thermostability tests, a drop in surface area was observed for all catalysts. In the case of the non-promoted COMON catalyst the surface area decreased insignificantly to 15.0 m^2/g. The loss was much more noticeable for COMON catalysts promoted with potassium compounds. For all of these catalysts, the specific surface area after aging decreased by an average of 1–2 m^2/g. For example, at a concentration of 1.5 wt.% of potassium, it decreased from about 9.5 to 8.3 m^2/g, and at a concentration of 2.9 wt.% of potassium it decreased from about 5.5 to 3.7 m^2/g. The most prominent decrease in the specific surface area, by about half, was observed for a catalyst containing 3.5 wt.% of potassium.

Catalytic activity measured at 450 °C during the ammonia synthesis reaction carried out under a pressure of 10 MPa for a series of COMON catalysts is shown in Figure 4. Considering fresh catalysts, the one containing 0.2 wt.% of potassium was less active than the non-promoted COMON catalyst. However, with increasing potassium concentration, the catalytic activity grew and its highest value was observed at 1.3% of potassium. In comparison to the non-promoted COMON catalyst, the activity at maximum was about 50% higher. The potassium concentration greater than 1.3% resulted in a drop in catalytic activity. Catalysts containing 2.9% and 3.5% potassium demonstrated very low activity, only about 25% and 10% compared to the non-promoted COMON catalyst, respectively.

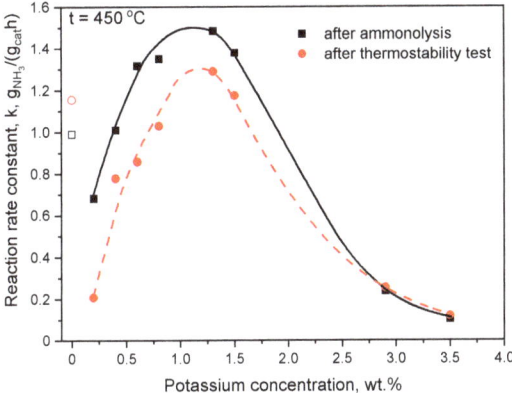

Figure 4. Catalytic activity of COMON catalysts as a function of the potassium concentration measured after ammonolysis and after the thermostability test. Catalytic tests were carried out at 450 °C under 10 MPa. Lines are given for eye guidance.

The second activity test was performed after the thermostability test. The catalytic activity of the non-promoted COMON catalyst increased by about 15%. After the thermostability test, significant deactivation was observed for most of the potassium-promoted COMON catalysts. The catalysts containing 1.3 wt.% and 1.5 wt.% of potassium were still more active than the reference material. However, a decrease of about 15% of their initial activity occurred. In the case of the catalysts containing 2.9 wt.% and 3.5 wt.% of potassium, which were barely active already in the initial stage, the catalytic activity remained virtually unchanged.

3. Discussion

The literature considering the early studies of cobalt molybdenum nitride catalysts indicates that the compound described by Co$_3$Mo$_3$N stoichiometry is expected to be the only ternary nitride obtained after ammonolysis of the precursor [15]. Metallic cobalt and molybdenum nitride, Mo$_2$N, were observed as intermediates and as decomposition products after prolonged gas and heat treatment [7,15,31]. However, in our previous studies [18–21] the occurrence of two cobalt molybdenum nitrides: Co$_3$Mo$_3$N and Co$_2$Mo$_3$N was observed.

Unlike previous reports by Kojima and Aika [24], the decomposition products (metallic cobalt or molybdenum nitride Mo_2N) were not observed after the thermostability test. Earlier reports indicate that the formation of the Co_2Mo_3N phase required special treatment [32]. However, a detailed review of older reports indicated that the Co_2Mo_3N phase also occurred after ammonolysis, but was not adequately identified [33].

In the study focused on the influence of chromium salts on the formation of similar mixtures of cobalt molybdenum nitrides [20], two possible ways of Co_2Mo_3N formation were considered. The Co_2Mo_3N phase was considered as a product of the decomposition of Co_3Mo_3N or as an intermediate on the path to Co_3Mo_3N formation. Since this phase disappears with increasing ammonolysis temperature, it is not observed in Co_3Mo_3N decomposition products either, it is supposedly an intermediate product. This is in agreement with our previous study [22], in which data obtained by the in situ XRD method during the non-promoted precursor activation process at 700 °C in the presence of a mixture of inert gas and ammonia was analyzed. Co_2Mo_3N is the first reaction product, which transforms into stable Co_3Mo_3N by further reconstruction with cobalt atoms.

In the present study, the preparation parameters: relatively low final ammonolysis temperature (700 °C) and high heating rate (10 °C/min) appear to promote the formation of the Co_2Mo_3N compound. Herein, a notable variation of the phase composition of the catalysts was observed. The precursors of the samples are virtually identical, and the change in Co_2Mo_3N concentration is attributed to the only variable parameter, the potassium concentration. The analogous effect of chromium salt addition was previously reported [20]. The content of the Co_2Mo_3N phase decreases with increasing concentration of chromium in the material. In the case of the potassium admixture, this dependence is more complex. Starting from small concentrations of potassium, the weight fraction of Co_2Mo_3N phase grows, with the maximum at about 0.8 wt.% of potassium. The excessive potassium content leads to a decrease in Co_2Mo_3N concentration. The mechanism that explains the observed influence of potassium on the phase composition of the studied samples remains unknown.

The catalytic activity of COMON catalysts in the process of ammonia synthesis is high. The highest reaction rate constant observed for the catalyst containing 1.3 wt.% of potassium was 1.5 $g_{NH_3} \cdot g^{-1} \cdot h^{-1}$. It is approximately twice as high as the reaction rate constant observed under identical process conditions for industrial iron catalysts (0.6 ÷ 0.7 $g_{NH_3} \cdot g^{-1} \cdot h^{-1}$) [34]. The optimal potassium concentration corresponds well to the studies we previously reported on COMON catalysts at 400 °C [18,27]. A comparable amount of potassium promoter was also claimed as optimal in the report by Kojima and Aika [15]. In their study, the catalyst containing approximately 1.2 wt.% of potassium (that is, 0.05 mol K per mol Mo) was the most active one. It must be stated that the latter study was carried out under lower pressures (between 0.1 MPa and 3.1 MPa).

The detrimental influence of excessive potassium admixture confirms earlier reports [15,18]. Especially beyond 2.9 wt.% of potassium, the catalytic activity of COMON catalysts was very low. An excess of alkali metal was supposed to prevent the proper development of the catalyst surface [15]. The surface area of cobalt molybdenum nitride mixtures was prominently affected by the potassium content. It decreased by about 30% compared to the potassium-free COMON catalyst. It also varied considerably with potassium concentration. The optimal potassium content in COMON catalysts needed for the development of the highest specific surface area is ambiguous. The values of this parameter observed at 0.8 wt.% and 1.3 wt.% are relatively close to each other, and the optimal potassium concentration supposedly lies between them.

The influence of the potassium content on the activity of the COMON catalysts is analogous to that observed for potassium-promoted iron catalysts, both fused [35] and supported [36]. Low and very high potassium concentrations in these catalysts also result in a relatively low catalytic activity. We suppose that this is a general property resulting from the presence of potassium atoms on the surface of the catalysts. Alkali metals are assumed to affect the electronic properties of the surface active sites. However, the possibility that potassium modifies the structure of catalysts should not be overlooked [37,38].

The loss of activity after the thermostability test correlates well with the decrease of the surface area, which was observed for all catalysts, apart from the unpromoted one. Because the only variable between the studied materials was the potassium concentration, this phenomenon is associated with the difference in the structure of the catalysts, which apparently is the change of Co_2Mo_3N and Co_3Mo_3N concentrations.

4. Materials and Methods

4.1. Precursor Synthesis

Catalyst precursors were obtained using the method described in the previous work [18]. Briefly, water solutions of cobalt nitrate, $Co(NO_3)_2 \cdot 6H_2O$, and ammonium molybdate, $(NH_4)_6Mo_7O_{24} \cdot 4H_2O$, were stirred and heated to about 90 °C. These solutions were mixed, while the pH of the resulting solution was controlled by addition of a 25% aqueous ammonia solution ($NH_3 \cdot H_2O$) to remain at pH = 5.5. A purple-blue precipitate was isolated by vacuum filtration, rinsed three times with distilled water and once with ethanol, then dried overnight at 150 °C. Potassium-promoted samples were obtained by impregnation of the precipitate in aqueous solutions of potassium nitrate, KNO_3, in a vacuum evaporator at 60 °C. The concentration of potassium ions in solution was chosen as such to obtain the potassium concentration in the final, nitrided form of catalysts in the range 0.2 – 3.5% by weight.

4.2. Nitriding of Precursor

The active form of the catalysts was synthesized via the reduction process of the oxidized precursor under the flow of pure ammonia in a horizontal steel reactor placed inside an electric oven and then following the procedure described elsewhere [31,39]. Approximately 6 g of the oxidized precursor powder was placed in a ceramic boat. After flushing the reactor with pure ammonia, the precursor was heated under flowing ammonia gas (NH_3 flow—250 sccm, heating rate—10 °C/min., maximum temperature—700 °C). The sample was kept under ammonia flow at 700 °C for 6 h and then cooled to room temperature. The resulting fine-crystalline substrate was pyrophoric, and therefore each sample was left overnight in the flow of oxygen/nitrogen mixture (1:100) for passivation. The materials were then removed from the reactor and pressed into pellets, which were subsequently crushed and sieved. The 1.0–1.2 mm grain fraction was selected and used for the activity experiments.

4.3. Material Characterization

The phase composition of the materials was analyzed by powder X-ray diffraction (XRD). The Philips X'pert PRO MPD diffractometer was used in Bragg–Brentano geometry, with a Cu radiation source. To avoid fluorescence effects, a graphite monochromator was used. Phase identification was performed with the use of the ICDD PDF-4+ database [40]. A full-pattern fit based on the Rietveld method, using the formalism described by Hill and Howard [41], was applied to calculate the weight fractions of the crystallographic phases identified in the material. A semi-automatic Rietveld refinement procedure included in the HighScore Plus software [42] by PANalytical B.V. was used. All the data required for initialization of the Rietveld refinement were retrieved from the ICDD database. During the Rietveld refinement, the scale factor, unit cell parameters, full width at half maximum, and peak shape parameters of the phases have been refined. The pseudo-Voigt function was used.

The specific surface area of the nitrided samples was measured by the volumetric method using the N_2 adsorption-desorption isotherm at 77 K. The Quantachrome Quadrasorb SI-Kr/MP apparatus was used. Before measurements, the samples were degassed in vacuo for 6 h at 400 °C. The specific surface area was calculated using the Brunauer–Emmett–Teller (BET) equation. It was performed using commercial QuadraWin software.

4.4. Catalytic Activity Tests

Catalytic activity tests were performed in the apparatus described in detail elsewhere [43]. The equipment consists of a 6-channel high-pressure steel reactor with a gas purification stage and enables the synthesis of ammonia under the pressure up to 10 MPa at a temperature reaching 650 °C. The 1 g samples of nitrided catalysts were placed inside the reactor in separate channels. Cobalt molybdenum nitride without alkali admixture was used as a reference sample. The samples were activated under a flowing reactant mixture ($N_2 + 3H_2$, 330 sccm, 0.1 MPa) according to the following temperature program: 2 h at 350 °C, 3 h at 400 °C, 14 h at 450 °C and 24 h at 500 °C. This procedure is intended to remove the superficial oxide layer formed on the surface of catalysts during the passivation stage. The ammonia synthesis process was carried out with parameters as follows: pressure—10 MPa, gas reactants flow—330 sccm, temperature—450 °C. The ammonia concentration was measured in the outlet gas stream using a Siemens ULTRAMAT 6 NDIR (non-dispersive infrared absorbance) gas analyzer. The reaction rate constants of the ammonia synthesis reaction was calculated for each catalyst utilizing the modified Tiemkin–Pyzhev equation described elsewhere [44].

After the first activity test, all catalysts were heated in the reactor under a flowing ammonia–hydrogen mixture at 650 °C for 12 h. This procedure was intended to simulate the long-run action of the catalyst and is further referred to as thermostability test. Subsequently, the temperature was lowered to 450 °C and the activity test was repeated under identical conditions as described above.

5. Conclusions

Catalysts based on the mixture of Co_3Mo_3N and Co_2Mo_3N phases are highly active in the process of ammonia synthesis. Admixture of potassium compounds promotes the catalytic activity. Additionally, the phase composition of the catalysts is affected by the potassium content. After the thermostability test, the potassium-free catalyst remains virtually unchanged. Potassium-promoted catalysts lose catalytic activity as a result of the decrease of their surface area. The characteristic shape of the relation between potassium concentration and all of the measured parameters was observed for fresh catalysts as well as after the thermostability test. The maximum of the surface area and activity was observed for the catalysts with the greatest concentration of the Co_2Mo_3N phase, around 0.8–1.3 wt.% of potassium.

Author Contributions: Conceptualization; methodology; investigation; writing—original draft preparation; writing—review and editing; visualization; project administration; funding acquisition, P.A. Investigation; writing—original draft preparation; writing—review and editing; visualization, W.C. Conceptualization: methodology; investigation; writing—original draft preparation; writing—review and editing; visualization; supervision, D.M. All authors have read and agreed to the published version of the manuscript.

Funding: The scientific work was financed by the Polish National Centre for Research and Development, grant "Lider", project No. LIDER/10/0039/L-10/18/NCBR/2019 (Paweł Adamski); and the National Science Centre, Poland, grant "Preludium Bis", project No. 2019/35/O/ST5/02500 (Wojciech Czerwonko).

Data Availability Statement: Not applicable.

Conflicts of Interest: The authors declare no conflict of interest.

References

1. Hargreaves, J.S.J. Nitrides as ammonia synthesis catalysts and as potential nitrogen transfer reagents. *Appl. Petrochem. Res.* **2014**, *4*, 3–10. [CrossRef]
2. Kotarba, A.; Dmytrzyk, J.; Raróg-Pilecka, W.; Kowalczyk, Z. Surface heterogeneity and ionization of Cs promoter in carbon-based ruthenium catalyst for ammonia synthesis. *Appl. Surf. Sci.* **2003**, *207*, 327–333. [CrossRef]
3. Karolewska, M.; Truszkiewicz, E.; Mierzwa, B.; Kępiński, L.; Raróg-Pilecka, W. Ammonia synthesis over cobalt catalysts doped with cerium and barium. Effect of the ceria loading. *Appl. Catal. A Gen.* **2012**, *445–446*, 280–286. [CrossRef]

4. Dongil, A.B. Recent Progress on Transition Metal Nitrides Nanoparticles as Heterogeneous Catalysts. *Nanomaterials* **2019**, *9*, 1111. [CrossRef]
5. Boisen, A.; Dahl, S.; Jacobsen, C.J.H. Promotion of Binary Nitride Catalysts: Isothermal N_2 Adsorption, Microkinetic Model, and Catalytic Ammonia Synthesis Activity. *J. Catal.* **2002**, *208*, 180–186. [CrossRef]
6. She, Y.; Tang, B.; Li, D.; Tang, X.; Qiu, J.; Shang, Z.; Hu, W. Mixed Nickel-Cobalt-Molybdenum Metal Oxide Nanosheet Arrays for Hybrid Supercapacitor Applications. *Coatings* **2018**, *8*, 340. [CrossRef]
7. Hada, K.; Tanabe, J.; Omi, S.; Nagai, M. Characterization of cobalt molybdenum nitrides for thiophene HDS by XRD, TEM, and XPS. *J. Catal.* **2002**, *207*, 10–22. [CrossRef]
8. Hada, K.; Nagai, M.; Omi, S. Characterization and HDS activity of cobalt molybdenum nitrides. *J. Phys. Chem. B* **2001**, *105*, 4084–4093. [CrossRef]
9. Logan, J.W.; Heiser, J.L.; McCrea, K.R.; Gates, B.D.; Bussell, M.E. Thiophene hydrodesulfurization over bimetallic and promoted nitride catalysts. *Catal. Lett.* **1998**, *56*, 165–171. [CrossRef]
10. Furimsky, E. Metal carbides and nitrides as potential catalysts for hydroprocessing. *Appl. Catal. A Gen.* **2003**, *240*, 1–28. [CrossRef]
11. Chen, X.; Zhang, T.; Zheng, M.; Wu, Z.; Wu, W.; Li, C. The reaction route and active site of catalytic decomposition of hydrazine over molybdenum nitride catalyst. *J. Catal.* **2004**, *224*, 473–478. [CrossRef]
12. Shi, C.; Zhu, A.M.; Yang, X.F.; Au, C.T. NO Reduction with Hydrogen over Cobalt Molybdenum Nitride and Molybdenum Nitride: A Comparison Study. *Catal. Lett.* **2004**, *97*, 9–16. [CrossRef]
13. Kojima, R.; Aika, K.I. Cobalt molybdenum bimetallic nitride catalysts for ammonia synthesis. Part 2. Kinetic study. *Appl. Catal. A* **2001**, *218*, 121–128. [CrossRef]
14. Jacobsen, C.J.H.; Dahl, S.; Clausen, B.S.; Bahn, S.; Logadottir, A.; Norskov, J.K. Catalyst Design by Interpolation in the Periodic Table: Bimetallic Ammonia Synthesis Catalysts. *J. Am. Chem. Soc.* **2001**, *123*, 8404–8405. [CrossRef] [PubMed]
15. Kojima, R.; Aika, K.-I. Cobalt molybdenum bimetallic nitride catalysts for ammonia synthesis Part 1. Preparation and characterization. *Appl. Catal. A Gen.* **2001**, *215*, 149–160. [CrossRef]
16. Jacobsen, C.J.H. Novel class of ammonia synthesis catalysts. *Chem. Commun.* **2000**, 1057–1058. [CrossRef]
17. Kojima, R.; Aika, K. Cobalt molybdenum bimetallic nitride catalysts for ammonia synthesis. *Chem. Lett.* **2000**, *29*, 514–515. [CrossRef]
18. Moszyński, D.; Jędrzejewski, R.; Ziebro, J.; Arabczyk, W. Surface and catalytic properties of potassium-modified cobalt molybdenum catalysts for ammonia synthesis. *Appl. Surf. Sci.* **2010**, *256*, 5581–5584. [CrossRef]
19. Adamski, P.; Nadziejko, M.; Komorowska, A.; Sarnecki, A.; Albrecht, A.; Moszyński, D. Chromium-modified cobalt molybdenum nitrides as catalysts for ammonia synthesis. *Open Chem.* **2019**, *17*, 127–131. [CrossRef]
20. Moszyński, D. Controlled phase composition of mixed cobalt molybdenum nitrides. *Int. J. Refract. Met. Hard Mater.* **2013**, *41*, 449–452. [CrossRef]
21. Moszyński, D.; Adamski, P.; Nadziejko, M.; Komorowska, A.; Sarnecki, A. Cobalt molybdenum nitrides co-promoted by chromium and potassium as catalysts for ammonia synthesis. *Chem. Pap.* **2018**, *72*, 425–430. [CrossRef]
22. Adamski, P.; Moszyński, D.; Komorowska, A.; Nadziejko, M.; Sarnecki, A.; Albrecht, A. Ammonolysis of Cobalt Molybdenum Oxides—In Situ XRD Study. *Inorg. Chem.* **2018**, *57*, 9844–9850. [CrossRef] [PubMed]
23. Liu, H. *Ammonia Synthesis Catalysts: Innovation and Practice*; World Scientific: Singapore, 2013.
24. Kojima, R.; Aika, K.I. Cobalt molybdenum bimetallic nitride catalysts for ammonia synthesis. Part 3. Reactant gas treatment. *Appl. Catal. A* **2001**, *219*, 157–170. [CrossRef]
25. Adamski, P.; Moszyński, D.; Nadziejko, M.; Komorowska, A.; Sarnecki, A.; Albrecht, A. Thermal stability of catalyst for ammonia synthesis based on cobalt molybdenum nitrides. *Chem. Pap.* **2019**, *73*, 851–859. [CrossRef]
26. Daisley, A.; Costley-Wood, L.; Hargreaves, J.S.J. The Role of Composition and Phase upon the Lattice Nitrogen Reactivity of Ternary Molybdenum Nitrides. *Top. Catal.* **2021**, *73*, 851–859. [CrossRef]
27. Nadziejko, M.; Adamski, P.; Moszyński, D. Doped cobalt-molybdenum catalysts for the ammonia synthesis. *Przem. Chem.* **2020**, *99*, 1454–1458. [CrossRef]
28. Moszyński, D.; Adamski, P.; Pelech, I.; Arabczyk, W. Cobalt-molybdenum catalysts doped with cesium for ammonia synthesis. *Przem. Chem.* **2015**, *94*, 1399–1403. [CrossRef]
29. Dry, M.E.; du Plessis, J.A.K.; Leuteritz, G.M. The Influence of Structural Promoters on the Surface Properties of Reduced Magnetite Catalysts. *J. Catal.* **1966**, *6*, 194–199. [CrossRef]
30. Raje, A.P.; O'Brien, R.J.; Davies, B.H. Effect of potassium promotion on iron-based catalysts for Fischer-Tropsch synthesis. *J. Catal.* **1998**, *180*, 36–43. [CrossRef]
31. Choi, J.-G.; Curl, R.L.; Thompson, L.T. Molybdenum Nitride Catalysts I. Influence of the Synthesis Factors on Structural Properties. *J. Catal.* **1994**, *146*, 218–227. [CrossRef]
32. Prior, T.J.; Battle, P.D. Facile synthesis of interstitial metal nitrides with the filled b-manganese structure. *J. Solid State Chem.* **2003**, *172*, 138–147. [CrossRef]
33. Alconchel, S.; Sapiña, F.; Beltran, D.; Beltran, A. Chemistry of interstitial molybdenum ternary nitrides M_nMo_3N (M = Fe, Co, n = 3; M = Ni, n = 2). *J. Mater. Chem.* **1998**, *8*, 1901–1909. [CrossRef]
34. Arabczyk, W.; Kałucki, K.; Kaleńczuk, R.J.; Śpiewak, Z.; Morawski, A.W.; Pajewski, R.; Ludwiczak, S.; Stołecki, K.; Janecki, Z. Badanie aktywności kontaktów żelazowych do syntezy amoniaku. *Przem. Chem.* **1986**, *65*, 532.

35. Kowalczyk, Z.; Jodzis, S.; Środa, J.; Diduszko, R.; Kowalczyk, E. Influence of aluminium and potassium on activity and texture of fused iron Catalysts for ammonia synthesis. *Appl. Catal. A Gen.* **1992**, *87*, 1–14. [CrossRef]
36. Yan, P.; Guo, W.; Liang, Z.; Meng, W.; Yin, Z.; Li, S.; Li, M.; Zhang, M.; Yan, J.; Xiao, D.; et al. Highly efficient K-Fe/C catalysts derived from metal-organic frameworks towards ammonia synthesis. *Nano Res.* **2019**, *12*, 2341–2347. [CrossRef]
37. Spencer, M. On the rate-determining step and the role of potassium in the catalytic synthesis of ammonia. *Catal. Lett.* **1992**, *13*, 45–53. [CrossRef]
38. Arabczyk, W.; Narkiewicz, U.; Moszyński, D. Double-layer model of the fused iron catalyst for ammonia synthesis. *Langmuir* **1999**, *15*, 5785–5789. [CrossRef]
39. Bem, D.S.; Olsen, H.P.; zur Loye, H.-C. Synthesis, Electronic and Magnetic Characterization of the Ternary Nitride ($Fe_{0.8}Mo_{0.2}$)MoN_2. *Chem. Mater.* **1995**, *7*, 1824. [CrossRef]
40. Gates-Rector, S.; Blanton, T. The Powder Diffraction File: A quality materials characterization database. *Powder Diffr.* **2019**, *34*, 352–360. [CrossRef]
41. Hill, R.; Howard, C. Quantitative phase analysis from neutron powder diffraction data using the Rietveld method. *J. Appl. Crystallogr.* **1987**, *20*, 467–474. [CrossRef]
42. Degen, T.; Sadki, M.; Bron, E.; König, U.; Nénert, G. The HighScore suite. *Powder Diffr.* **2014**, *29*, S13–S18. [CrossRef]
43. Arabczyk, W. The state of studies on iron catalyst for ammonia synthesis. *Pol. J. Chem. Technol.* **2005**, *7*, 8–17.
44. Arabczyk, W.; Moszyński, D.; Narkiewicz, U.; Pelka, R.; Podsiadły, M. Poisoning of iron catalyst by sulfur. *Catal. Today* **2007**, *124*, 43–48. [CrossRef]

Article

Physicochemical Features and NH₃-SCR Catalytic Performance of Natural Zeolite Modified with Iron—The Effect of Fe Loading

Magdalena Saramok [1],*, Marek Inger [1],*, Katarzyna Antoniak-Jurak [1], Agnieszka Szymaszek-Wawryca [2], Bogdan Samojeden [2] and Monika Motak [2]

[1] Łukasiewicz Research Network—New Chemical Syntheses Institute, Al. Tysiąclecia Państwa Polskiego 13a, 24-110 Puławy, Poland; katarzyna.antoniak@ins.lukasiewicz.gov.pl

[2] Faculty of Energy and Fuels, AGH University of Science and Technology, Al. Mickiewicza 30, 30-059 Kraków, Poland; agnszym@agh.edu.pl (A.S.-W.); bsamo1@agh.edu.pl (B.S.); motakm@agh.edu.pl (M.M.)

* Correspondence: magdalena.saramok@ins.lukasiewicz.gov.pl (M.S.); marek.inger@ins.lukasiewicz.gov.pl (M.I.)

Abstract: In modern dual-pressure nitric acid plants, the tail gas temperature usually exceeds 300 °C. The NH₃-SCR catalyst used in this temperature range must be resistant to thermal deactivation, so commercial vanadium-based systems, such as V_2O_5-WO_3 (MoO_3)-TiO_2, are most commonly used. However, selectivity of this material significantly decreases above 350 °C due to the increase in the rate of side reactions, such as oxidation of ammonia to NO and formation of N_2O. Moreover, vanadium compounds are toxic for the environment. Thus, management of the used catalyst is complicated. One of the alternatives to commercial V_2O_5-TiO_2 catalysts are natural zeolites. These materials are abundant in the environment and are thus relatively cheap and easily accessible. Therefore, the aim of the study was to design a novel iron-modified zeolite catalyst for the reduction of NO_x emission from dual-pressure nitric acid plants via NH₃-SCR. The aim of the study was to determine the influence of iron loading in the natural zeolite-supported catalyst on its catalytic performance in NO_x conversion. The investigated support was firstly formed into pellets and then impregnated with various contents of Fe precursor. Physicochemical characteristics of the catalyst were determined by XRF, XRD, low-temperature N_2 sorption, FT-IR, and UV–Vis. The catalytic performance of the catalyst formed into pellets was tested on a laboratory scale within the range of 250–450 °C using tail gases from a pilot nitric acid plant. The results of this study indicated that the presence of various iron species, including natural isolated Fe^{3+} and the introduced FexOy oligomers, contributed to efficient NO_x reduction, especially in the high-temperature range, where the NO_x conversion rate exceeded 90%.

Keywords: nitric acid plant; selective catalytic reduction; clinoptilolite; iron-modified zeolite catalyst

Citation: Saramok, M.; Inger, M.; Antoniak-Jurak, K.; Szymaszek-Wawryca, A.; Samojeden, B.; Motak, M. Physicochemical Features and NH₃-SCR Catalytic Performance of Natural Zeolite Modified with Iron—The Effect of Fe Loading. *Catalysts* **2022**, *12*, 731. https://doi.org/10.3390/catal12070731

Academic Editor: Anker Degn Jensen

Received: 18 May 2022
Accepted: 28 June 2022
Published: 1 July 2022

Publisher's Note: MDPI stays neutral with regard to jurisdictional claims in published maps and institutional affiliations.

Copyright: © 2022 by the authors. Licensee MDPI, Basel, Switzerland. This article is an open access article distributed under the terms and conditions of the Creative Commons Attribution (CC BY) license (https://creativecommons.org/licenses/by/4.0/).

1. Introduction

NO_x emitted from stationary (power plants, nitric acid, or adipic acid production) and mobile sources are treated as a serious environmental problem. They contribute to the formation of acid rain and photochemical smog and cause deterioration of water and soil quality [1]. Therefore, it is highly necessary to reduce industrial NO_x emissions. The method of NO_x abatement is usually correlated with the emission origin. In the plants, which produce nitric acid, high-efficiency absorption, non-selective catalytic reduction (NSCR), selective catalytic reduction (SCR), and absorption in sodium hydroxide solution can be used [2]. Among them, selective catalytic reduction with ammonia (NH₃-SCR) is the most efficient. The process involves selective reduction of NO_x with NH_3 to form N_2 and H_2O, as presented by Equations (1)–(3):

$$4\,NO + 4\,NH_3 + O_2 \rightarrow 4\,N_2 + 6\,H_2O \tag{1}$$

$$2\,NO_2 + 4\,NH_3 + O_2 \rightarrow 3\,N_2 + 6\,H_2O \tag{2}$$

$$NO_2 + NO + 2\,NH_3 \rightarrow 2\,N_2 + 3\,H_2O \tag{3}$$

Ammonia, used as the reducing agent, is easily available in nitric acid plants since it is a substrate in the production of HNO_3. Typically, the SCR reactor is installed at the end of the technological line and does not significantly affect the production of acid. Therefore, NH_3-SCR can be used in most existing nitric acid plants. The catalyst used in the NH_3-SCR process is required to exhibit high activity in low- and high-temperature regions, satisfactory selectivity to N_2, and good thermal stability. In fact, these requirements are met by metal oxide systems, such as the commercial catalyst V_2O_5-WO_3 (MoO_3)-TiO_2 [3]. However, the material is not free from some important drawbacks, such as the toxicity of vanadium compounds. Moreover, selectivity of the catalysts above 350 °C is limited by the side reactions described by Equations (4) and (5):

$$2\,NH_3 + 2\,O_2 \rightarrow N_2O + 3\,H_2O \tag{4}$$

$$4\,NH_3 + 5\,O_2 \rightarrow 4\,NO + 6\,H_2O \tag{5}$$

Due to the above-mentioned problems, a number of materials have been investigated as the alternative catalysts of NH_3-SCR [4–7]. According to the study reported by Kobayashi et al. [8], application of TiO_2 does not provide sufficient dispersion of the active phase, surface acidity, and thermal stability of the catalyst. Therefore, further research has shifted to alternative supports of the novel catalyst. Among them, natural zeolites were found to be very promising precursors of the novel catalysts [9,10]. The great advantage of these materials is their abundance in the environment and thus their relatively low price, which is very beneficial for industrial applications. The representative of natural zeolites is clinoptilolite, belonging to the heulandite (HEU) family [11]. The material shows strongly acidic character, determined by its Si/Al molar ratio of ca. 4, its well-developed pore system, and its thermal stability. According to the Eley–Rideal mechanism, NH_3-SCR assumes simultaneous adsorption of alkaline NH_3 and neutral NO and their interaction on the catalyst surface [12]. Therefore, high concentration of acid centers delivered by clinoptilolite improves ammonia adsorption capacity and NH_3-SCR reaction rate. Moreover, clinoptilolite provides good ion exchange capacity, and as a consequence, its acidic character can be easily elevated by acid pretreatment [13]. Additionally, the presence of micro- and mesopores in the zeolitic structure facilitates the diffusion of gas molecules through the catalyst's pores and easy access to active centers. Lastly, clinoptilolite belongs to the residual materials, usually stored on heaps. Therefore, its recycling is in agreement with the assumptions of circular economy. All in all, the above-mentioned properties make clinoptilolite a promising candidate for the precursor of a new catalyst of NH_3-SCR [13–15]. To date, research on the application of clinoptilolite was mostly limited to SCR with hydrocarbons as reducing agents. Ghasemian et al. [16] proved that protonated clinoptilolite is a promising precursor of a new catalyst of SCR with methane as a reducing agent. Another study conducted by the authors [15,17] concerned clinoptilolite as a possible support for the catalyst of SCR with propane. However, only few studies have explored zeolite as a support for SCR with ammonia [13,18].

Another important issue in the design of a novel NH_3-SCR catalyst is the active phase. Over recent years, the focus of researchers has shifted to systems with transition metals, especially iron [5,12,19,20]. The choice of Fe was motivated by its environmentally benign characteristics, low price, and prominent thermal stability. Additionally, iron catalysts exhibit excellent medium- and high-temperature activity and satisfactory selectivity to N_2.

Moreover, the facile redox equilibrium, $Fe^{3+} \leftrightarrow Fe_3O_4 \leftrightarrow Fe^{2+}$, contributes to high oxygen storage capacity, which is very beneficial for SCR catalysts.

Highly satisfactory activity of iron-modified clinoptilolite in SCR with ammonia was confirmed in the previous study [18]. It was found that raw clinoptilolite in a form of fine grain showed 30% of NO conversion in the range of 350–450 °C. The high efficiency of the material in NH_3-SCR with the gas mixture reflecting the industrial composition was also confirmed. It was observed that at 400 °C, NO_x conversion for Fe-clinoptilolite exceeded 80%, and high selectivity to N_2 was preserved in the entire temperature range. Additionally, 82% NO conversion was obtained for the previously shaped iron-modified clinoptilolite. In conditions similar to industrial ones, the highest catalytic activity was obtained above 400 °C, and these temperatures also maintained very favorable selectivity towards N_2. Importantly, no formation of N_2O was observed during the catalytic reaction.

In this work, the aim was to investigate the influence of iron loading on the low- and high-temperature catalytic performance of Fe-modified clinoptilolite formed into pellets. In this research, iron was considered the active phase since this transition metal exhibits outstanding redox properties and, at the same time, neutrality to the environment. Therefore, the experiments will contribute to the development of more ecologically friendly catalysts of the NH_3-SCR process. Moreover, in the experiments, a real tail gas mixture, which normally enters SCR reactors in nitric acid plants, was used. To the best of our knowledge, no one so far has investigated the catalytic performance of such material under near-industrial conditions. Thus, this work makes a significant contribution to the field of low-price and nontoxic industrial catalysts of NH_3-SCR.

2. Results and Discussion

2.1. Physicochemical Properties of the Materials

2.1.1. Chemical Composition, Crystal Structure, and Morphology of the Materials

The chemical compositions of raw (Clin), protonated (H-Clin), and Fe-modified clinoptilolite (Fe-Clin-1, 2, or 3) are presented in Table 1. The crystalline structure of the materials was analyzed using XRD, and the obtained patterns are shown in Figure 1.

Table 1. Chemical composition (in wt.%) of the analyzed materials determined by XRF.

Sample	Fe_2O_3 (%)	SiO_2 (%)	Al_2O_3 (%)	Na_2O (%)	MgO (%)	SO_3 (%)	K_2O (%)	CaO (%)	TiO (%)	MnO (%)
Clin	2.1	74.7	12.1	0.8	0.8	0.04	3.2	3.7	0.2	0.07
H-Clin	2.0	80.5	11.4	0.1	0.7	0.03	3.0	1.7	0.2	0.03
Fe-Clin-1	8.6	71.0	10.1	0.1	0.6	5.00	2.6	1.5	0.2	0.01
Fe-Clin-2	11.1	69.2	10.0	0.1	0.5	4.59	2.5	1.5	0.2	0.01
Fe-Clin-3	11.9	66.7	9.4	0.1	0.5	7.41	2.3	1.2	0.2	0.01

As presented in Table 1, the raw clinoptilolite consisted mainly of SiO_2 and Al_2O_3 and contained some additives of other alkaline metal oxides. Additionally, the analysis provided strong evidence of the presence of iron oxide in the natural zeolite. After protonation, the percentage contribution of SiO_2 increased with a simultaneous slight decrease of Al_2O_3 content. This result is in line with that obtained by Burris and Juenger [21], who ascribed the decrease in aluminum content to partial dealumination of the material or its dissolution in acidic medium. However, since the XRD pattern of H-Clin corresponded to that of Clin, the degrading influence of the acid can be excluded. After the deposition of iron, the detected content of Fe_2O_3 significantly increased, proving efficient incorporation of various iron species into the zeolite structure. However, it can be also observed that after the third impregnation, the amount of Fe_2O_3 was very close to that obtained after the second impregnation. Additionally, the catalysts contained considerable amounts of SO_3 as the result of using $FeSO_4$ as the precursor of iron. Therefore, the applied calcination temperature was probably insufficient to provide effective decomposition of the salt deposited on the zeolite matrix.

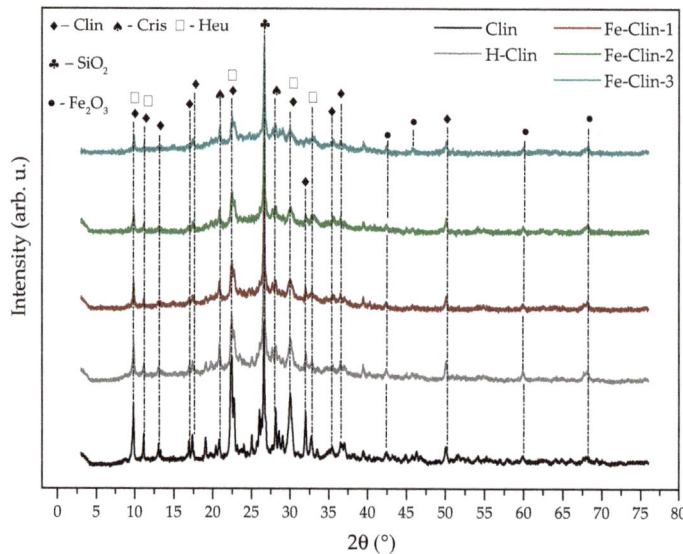

Figure 1. XRD patterns obtained for raw clinoptilolite (Clin), protonated clinoptilolite (H-Clin), and clinoptilolite modified with iron (Fe-Clin-1, Fe-Clin-2, Fe-Clin-3).

According to the results of XRD, the analyzed sample consisted mainly of heulandite/clinoptilolite, confirmed by the diffraction maxima at 2θ of 9.8, 11.4, 12.9, 16.8, 17.4, 20.7, 22.6, 30.0, 32.0, 32.9, 35.5, 36.7, and 50.3°. The reflection at 2θ of 26.6° corresponds to SiO_2, while those at 21.9 and 28.1° are due to the presence of cristobalite impurities in the solid [13]. The observed diffraction maxima are in good agreement with those reported in the literature [18,22]. The comparative analysis of the materials showed that protonation by acid treatment did not result in any noticeable structural changes. However, some of the diffraction maxima exhibited lower intensity or completely disappeared, indicating decreased crystallinity of the catalysts compared to the raw zeolite.

After modification with iron, the positions of diffraction maxima characteristic of the clinoptilolite phase remained unchanged. Thus, deposition of the active phase did not cause any significant damage to the structure. However, the intensity of the reflections was the lowest for the material with the highest concentration of iron. Larger aggregates of Fe_2O_3 on the zeolite surface can potentially be present at 2θ of 42.0, 45.8, 60.2, and 68.2° [23]. However, apart from bigger particles, iron also isomorphously substituted for aluminum in the zeolite framework and thus was impossible to be detected by XRD technique. The replacement of Al by Fe can be also confirmed by lower intensities of the structural diffraction maxima of clinoptilolite. A similar effect was obtained by Kessouri et al. [24] after the deposition of iron into an MFI framework. Hence, the noticeably decreased intensity of the reflections for Fe-Clin-3 can be explained by the highest rate of isomorphous substitution or deposition of bulky species of Fe_2O_3 on its surface.

2.1.2. Textural Properties of the Materials

Low-temperature N_2 adsorption–desorption isotherms obtained for raw clinoptilolite (Clin), protonated clinoptilolite (H-Clin), and iron-modified zeolite (Fe-Clin-1, Fe-Clin-2, Fe-Clin-3) are presented in Figure 2. Furthermore, pore volume distribution is shown in Figure 3, while the textural and structural parameters of the samples are summarized in Table 2. Raw clinoptilolite demonstrated IV(a) type isotherm with the hysteresis loop H3, according to the IUPAC classification [25]. This isotherm is characteristic of materials with wedge-shaped mesopores and nonrigid aggregates of platelike particles [25,26]. The specific surface area of the nonmodified clinoptilolite is in range of 16–30 $m^2 \cdot g^{-1}$, which is typical for clinoptilolite [27]. Vassileva and Voikova [26] reported that the relatively low values of specific surface area and pore volume exhibited by nonmodified clinoptilolite are caused by the limited access of N_2 molecules to the internal structure of the zeolite. As a result, the adsorbate was deposited mainly on the external surface of the material. The results of the experiment performed after NH_3-SCR tests showed that the specific surface area was preserved, even in the case of the catalytic reaction being conducted under severe conditions. After the dealumination procedure, the volume of mesopores in clinoptilolite significantly increased, suggesting the formation of a mesopore system. The isotherms obtained for iron-modified clinoptilolite are characterized by the isotherms of type IV(a), confirming their mesoporous nature. However, the introduction of iron resulted in a change in the shape of the hysteresis loop from H3 to H4 [28]. This result suggests the transformation of wedge-shaped mesopores into slit-shaped ones. Additionally, as presented in Table 2, after modification with iron, the specific surface area, the volume of mesopores, and the average pore diameter decreased due to pore blockage probably caused by the deposition of iron oxide species [29]. Catalysts of Fe-Clin-X series characteristically possess similar pore distribution (Figure 3). For Fe-Clin-X series, a wide bimodal pore distribution—pores with diameters ranging from 10 to 1000 Å and pores with diameters from 1500 to 7000 Å—was observed. However, the porous structure was definitely dominated by pores with diameters ranging from 10 to 1000 Å (mesopores with D_{meso} in the range of 208–223 Å). Interestingly, multiple impregnations with $FeSO_4$ did not result in considerable differences between the D_{meso} values. Nevertheless, the gradual decline of V_{meso} with the increasing iron content suggested that iron species were effectively deposited in the inner structure of clinoptilolite.

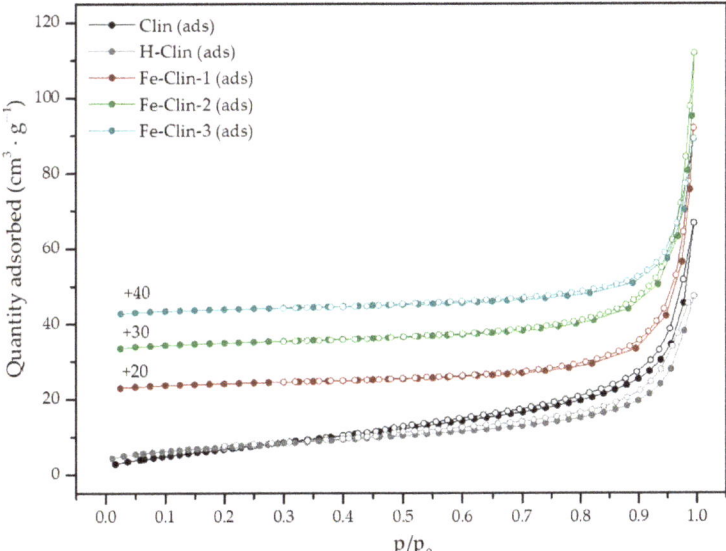

Figure 2. Low-temperature N_2 adsorption–desorption isotherms obtained for raw clinoptilolite (Clin) and the investigated catalysts (for better visibility, the isotherms were shifted by the values given in the figure).

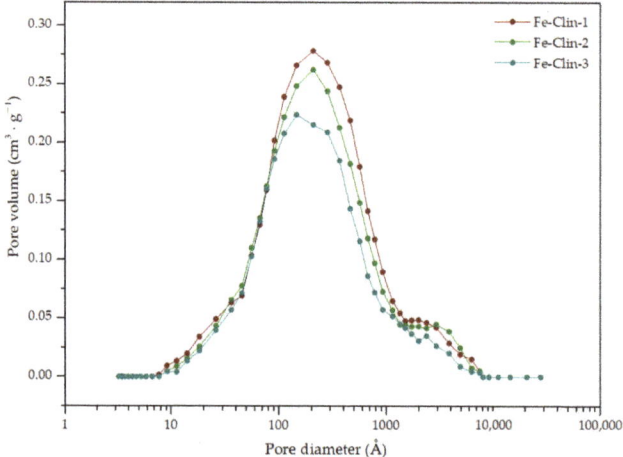

Figure 3. Pore volume obtained for Fe-clinoptilolite catalysts.

Table 2. Textural and structural parameters of raw clinoptilolite and the Fe-clinoptilolite catalysts.

Sample	S_{BET} [a] $(m^2 \cdot g^{-1})$	S_{Ext} [b] $(m^2 \cdot g^{-1})$	V_{meso} [c] $(cm^3 \cdot g^{-1})$	D_{meso} [c] (nm)
Clin	16	7	0.025	17.0
H-Clin	30	9	0.281	28.2
Fe-Clin-1	15	10	0.277	20.8
Fe-Clin-2	12	13	0.262	20.8
Fe-Clin-3	10	10	0.224	22.3

[a] Specific surface area determined using the BET method; [b] external surface area determined using the t-plot method; [c] average mesopore volume and diameter determined using the BJH method.

2.1.3. Characteristic Chemical Groups in the Materials

The FT-IR spectra obtained for the raw and protonated clinoptilolite and the zeolites with various loadings of iron are presented in Figure 4. The characteristic peaks can be divided into three regions: (1) O-H stretching vibrations (3800–3400 cm^{-1}); (2) Si-O stretching vibrations, Al-Me-OH stretching vibrations, and O-H bending vibrations from H$_2$O (1700–700 cm^{-1}); and (3) pseudo lattice vibrations (700–450 cm^{-1}) [13].

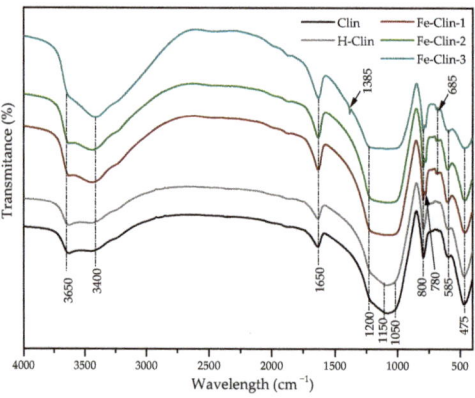

Figure 4. FT-IR spectra of raw clinoptilolite (Clin), protonated clinoptilolite (H-Clin), and clinoptilolite modified with different Fe contents (Fe-Clin-1, Fe-Clin-2, Fe-Clin-3).

In the first of the above-mentioned regions, 3800–3400 cm^{-1}, the shape of the spectra was similar for all of the materials except Fe-Clin-3. The peak at 3650 cm^{-1} suggested the presence of Brönsted sites provided by the acidic hydroxyl Si-O(H)-Al. In the case of Fe-Clin-3, it was not as sharp as for the other materials; thus, multiple repetition of Fe deposition resulted in the removal of OH groups bonded to the zeolitic structure. The band at 3400 cm^{-1}, broad for raw and protonated clinoptilolite and sharper for Fe-modified zeolite, corresponded to the vibrations of O-H$\cdots$O bonds [30].

In the second region of the spectra, 1700–700 cm^{-1}, the peak at 1650 cm^{-1}, attributed to deformation vibrations of physisorbed water molecules, showed a similar shape for all materials [31]. However, the characteristic bands at 1200 cm^{-1} and 1050 cm^{-1} were almost absent for clinoptilolite modified with Fe. Both of the peaks were related to Al-O or Si-O asymmetric stretching vibrations; thus, incorporation of iron resulted in structural interruptions, such as the removal of charge-balancing Ca^{2+} and Mg^{2+}. A similar effect was observed by Cobzaru et al. [32] after the modification of natural clinoptilolite with nitric acid. Since $FeSO_4$ is regarded as a strongly acidic medium, our results are in line with this research. Moreover, the band at 1150 cm^{-1}, intense for Clin and H-Clin, partially disappeared after the introduction of iron. Since the peak corresponds to three-dimensional networks of amorphous Si-O-Si units, the modification procedure could partially remove this phase from natural and protonated zeolite. Additionally, a small peak at 1385 cm^{-1}, appearing only for Fe-Clin-3 and thus with the highest concentration of iron species, was probably ascribed to sulfate groups bonded to iron ions deposited in the zeolitic structure [33].

The characteristic peaks detected in the third analyzed region, 700–450 cm^{-1}, evidenced partial removal of amorphous silica. This effect can be confirmed by the presence of the sharp peaks at 800 cm^{-1} in the spectra of all the materials. However, after modification with iron, these peaks were noticeably separated. The new small peak at 780 cm^{-1}, formed through this division, raised from the stretching vibrations of $[SiO_4]$ tetrahedra from the zeolitic framework. Therefore, removal of the amorphous silica could enhance the detection of structural peaks of the materials. Another difference in the spectra of iron-modified clinoptilolite compared to the raw or protonated form was the presence of low-intense peaks at 585 cm^{-1}, which corresponded to the symmetric stretching vibrations of $[AlO_4]$ tetrahedra [34]. Two bands at 600 cm^{-1} and 475 cm^{-1} were related to O-Al-O or O-Si-O bending vibrations and Si-O stretching vibrations, respectively [35].

2.1.4. Speciation of the Active Phase

The comparative UV–Vis spectra of the raw clinoptilolite and the catalysts with various contents of iron are presented in Figure 5.

In general, for iron-modified zeolites, three main regions in UV–Vis spectra are expected: (1) bands below 300 nm, corresponding to the oxygen-to-metal charge transfer (CT), assigned to isolated framework and extra framework pseudotetrahedral Fe^{3+} species; (2) bands in the range of 300–500 nm, related to oligomeric Fe_xO_y or Fe_2O_3 nanoparticles; and (3) bands detected above 500 nm, assigned to Fe_2O_3 clusters on the external surface of the support [19].

The speciation of iron in the analyzed materials was strongly correlated with the metal loading. As presented in Figure 5, all the investigated samples, including nonmodified clinoptilolite, showed absorption bands at 230 and 260 nm. This result confirmed that iron was originally present (Clin) or isomorphously deposited (Fe-Clin-1, 2, and 3) in the zeolite structure in the form of extraframework cations with octahedral coordination [36]. Furthermore, the band at 350 nm, detected only for the zeolite modified with iron, corresponded to small, oligonuclear clusters of iron oxide [37]. The bands at 475 nm, characteristic of bigger particles of Fe_2O_3, were observed for the samples with increased iron content (Fe-Clin-2, Fe-Clin-3). Therefore, the extraframework phase of Fe_2O_3 became dominant as a result of the increase in iron loading due to the agglomeration of the species into bigger particles.

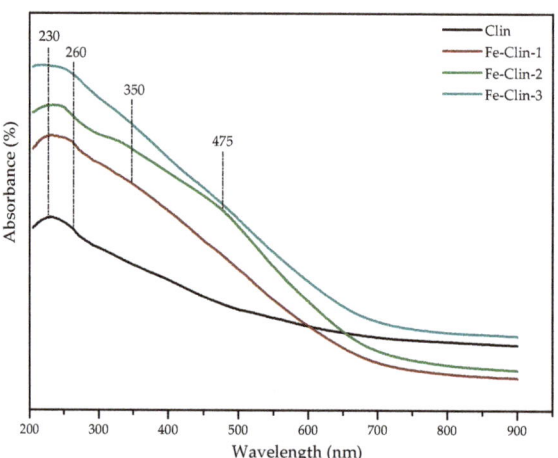

Figure 5. UV–Vis spectra of raw clinoptilolite (Clin) and clinoptilolite modified with different Fe contents (Fe-Clin-1, Fe-Clin-2, Fe-Clin-3).

2.2. NH$_3$-SCR Catalytic Tests Performed with Industrial Gas Mixture

NH$_3$-SCR catalytic tests over protonated clinoptilolite were conducted under the conditions reflecting that of industrial nitric acid plant (regarding catalytic bed loading and temperature range). The obtained results are presented in Figure 6. The tests were carried out at two catalytic bed loads. It was observed that in both cases, NO$_x$ conversion exceeded 50% in the entire temperature range. The maximum conversion of more than 90%, was reached above 400 °C, and higher NO$_x$ conversion was achieved for the lower catalytic bed loading. In the case of GHSV = 4500 h^{-1} (tail gas flow 0.15 Nm$^3 \cdot$h^{-1}), 93% of NO$_x$ conversion was obtained at 400 °C. On the other hand, for GHSV = 9000 h^{-1} (tail gas flow 0.3 Nm$^3 \cdot$h^{-1}), the material exhibited 80% of NO$_x$ conversion at 450 °C.

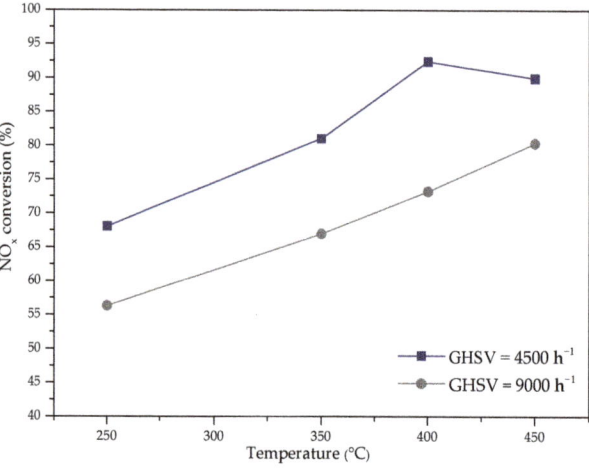

Figure 6. NO$_x$ conversion as a function of temperature and catalyst load for H-Clin.

During the test, N$_2$O concentration upstream and downstream of the catalytic bed was measured as well. The dash line in Figure 7 represents the ratio of the N$_2$O concentration downstream to the inlet concentration of N$_2$O. It was clearly indicated that higher loading of the catalytic bed resulted in lower N$_2$O concentration downstream of the bed compared to the inlet concentration of N_2O. This effect was observed over almost the entire investigated temperature range. Moreover, regardless of the catalytic bed load, the highest selectivity was observed at 450 °C.

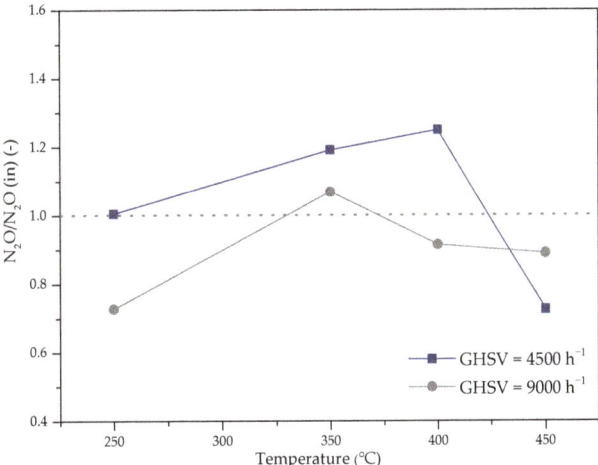

Figure 7. $N_2O/N_2O_{(in)}$ ratio as a function of temperature and catalyst load for H-Clin.

Figure 8 shows the results of the catalytic tests obtained for iron-modified samples and the protonated clinoptilolite, while the selectivity of the materials to N_2 is listed in Table 3. In all cases, NO_x conversion of iron-modified zeolite was higher than that of H-Clin. Above 350 °C, regardless of the iron content in the sample, NO_x conversion of over 90% was achieved. The highest activity in the entire temperature range was exhibited by Fe-Clin-2. Additionally, selectivity of Fe-clinoptilolite catalysts to N_2 was in the range of 93–100%, confirming the negligible contribution of the side reactions to the whole mechanism of NH_3-SCR performed on the materials.

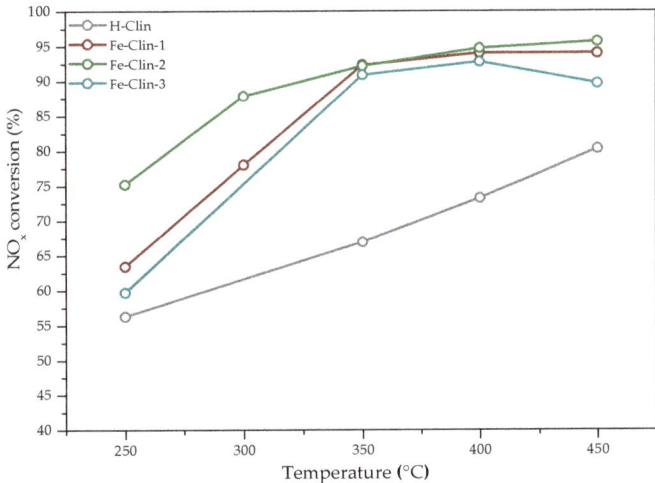

Figure 8. NO_x conversion as a function of temperature for protonated clinoptilolite (H-Clin) and Fe-clinoptilolite catalysts (Fe-Clin-1, Fe-Clin-2, Fe-Clin-3) at GHSV = 9000 h^{-1}.

Table 3. N_2 selectivity of Fe-clinoptilolite catalysts.

Sample	Selectivity Towards N_2 (%)				
	250 °C	300 °C	350 °C	400 °C	450 °C
Fe-Clin-1	99.2	98.2	96.7	96.9	100.0
Fe-Clin-2	99.6	98.6	97.7	94.6	100.0
Fe-Clin-3	98.3	-	97.1	93.1	100.0

The N_2O concentrations measured during the experiments are shown in Figure 9. In the case of protonated clinoptilolite, the N_2O concentration increased above the inlet value only at 350 °C. For iron-modified samples, the courses of the curves are similar to each other. Up to the temperature of 400 °C, the concentration of N_2O behind the bed slightly increased in relation to the initial concentration ($N_2O/N_2O_{(in)}$ > 1), and then, a sharp decrease in the concentration of N_2O at the temperature of 450 °C was noted. The greatest decrease was obtained for the Fe-Clin-1 and Fe-Clin-2 samples. Overall, satisfactory catalytic performance exhibited by the investigated catalysts confirmed that one or two iron impregnations of clinoptilolite are sufficient to obtain an effective NH_3-SCR catalyst.

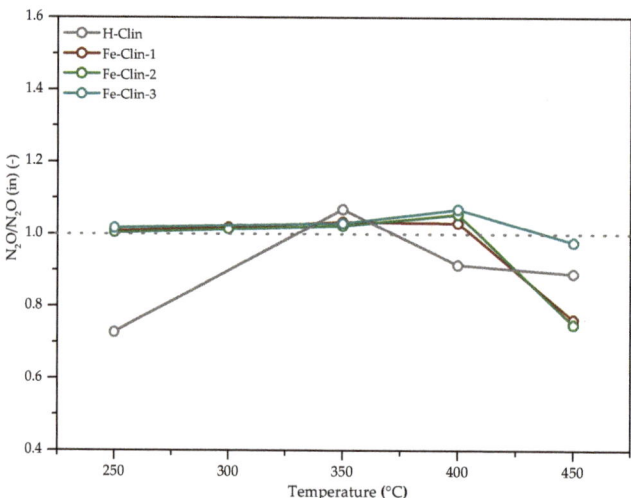

Figure 9. The ratio of N_2O concentration downstream of the catalytic bed and N_2O inlet concentration as a function of the temperature for H-Clin, Fe-Clin-1, Fe-Clin-2, Fe-Clin-3 at GHSV = 9000 h^{-1}.

3. Materials and Methods

3.1. Catalysts Preparation

The precursor of the investigated catalysts was raw zeolite with a high content of clinoptilolite phase. Firstly, the material was dealuminated using 5% HNO_3 solution. The operation was repeated three times in order to increase the dealumination rate. After each dealumination step, the precursor was washed with demineralized water until pH was <6 and dried at 105–110 °C. Subsequently, the zeolite was fractioned into 0.3–0.8 mm grains and formed into pellets of 5.0 × 4.8 mm dimensions, illustrated in Figure 10. Afterwards, the materials were calcined at 450 °C for 2 h. Iron-modified materials were prepared using the wet impregnation method using an aqueous solution of 1 M $FeSO_4$ as Fe precursor. The samples were left in contact with the solution at 50 °C for 1 h, then dried at 105–110 °C and calcined at 500 °C for 2 h. The impregnation procedure was performed one, two, or three times in order to obtained catalysts with various Fe loadings The precursors were dried and calcined before each impregnation treatment. The preparation procedure is schematically illustrated in Figure 10.

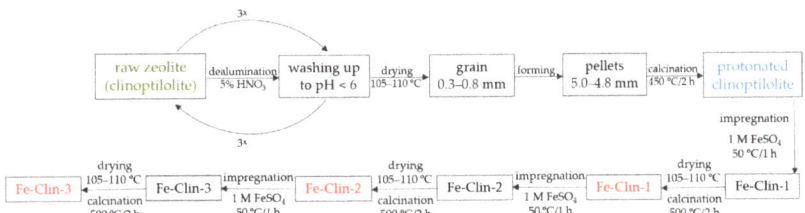

Figure 10. Schematic preparation procedure of Fe-clinoptilolite catalysts.

The formed samples of protonated clinoptilolite and Fe-clinoptilolite catalysts, prepared on a laboratory scale, are presented in Figure 11A,B, respectively. The codes of the samples with the corresponding descriptions are listed in Table 4.

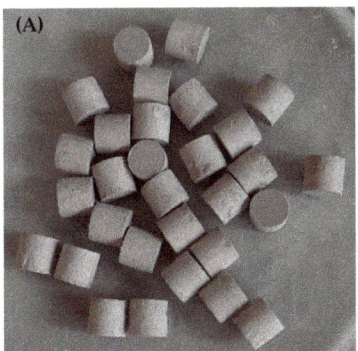

Figure 11. Protonated clinoptilolite (**A**) and Fe-clinoptilolite catalyst formed into pellets (**B**).

Table 4. The list of the samples with their codes and descriptions.

Sample	Description of the Sample
Clin	Raw clinoptilolite
H-Clin	Protonated clinoptilolite
Fe-Clin-1	The catalyst obtained by single impregnation with Fe precursor
Fe-Clin-2	The catalyst obtained by dual impregnation with Fe precursor
Fe-Clin-3	The catalyst obtained by triple impregnation with Fe precursor

3.2. Catalysts Characterization

X-ray fluorescence (XRF) was used to determine the chemical composition of the samples using Energy Dispersive X-ray Fluorescence EDXRF Spectrometer, Epsilon 3XLE PANalytical Company. The crystalline structure of the samples was analyzed using an X-ray diffraction (XRD) technique. X-ray diffraction patterns were obtained using an Empyrean diffractometer (Panalytical) equipped with a copper-based anode (Cu-Kα LFF HR, λ = 0.154059 nm). The measurement was conducted in the 2θ range of 2.0–70.0° (2θ step scans of 0.02° and the counting time of 1 s per step). The specific surface area, total pore volume, and mesopore volume were determined using an ASAP® 2050 Xtended Pressure sorption analyzer (Micromeritics Instrument Co., Norcross, GA, USA) based on N_2 adsorption–desorption isotherms at −196 °C using the BET adsorption model (Brunauer–Emmett–Teller) and the BJH transformation (Barret–Joyner–Halenda). Fourier transform infrared spectroscopy studies (FT-IR) were conducted using a Perkin Elmer Frontier FT-IR spectrometer. The spectra were obtained in the wavelength range of 4000–400 cm^{-1} with a resolution of 4 cm^{-1}. Before each measurement, the sample was mixed with KBr in a ratio of 1:100 and pressed into a disk. Coordination and aggregation of iron species were determined by UV–Vis spectroscopy at a wavelength range of 200–900 nm with a resolution of 1 nm using a Perkin Elmer Lambda 35 UV–Vis spectrophotometer.

3.3. Catalytic Tests in Real Gas Conditions

The activity and selectivity of the catalysts in the NH_3-SCR process were tested in the laboratory installation in the flow of the tail gases stream derived from the pilot ammonia oxidation plant. The laboratory installation consisted of a reactor (R) with a diameter of 25 mm and heat exchangers (HEx and HExNH$_3$) used for preheating tail gases and ammonia. In the tests, the height of the catalyst layer was 70 mm. The installation is schematically presented in Figure 12. The heated tail gases were mixed with ammonia and turned into the catalytic bed. The composition of the tail gases was similar to the tail gases emitted from industrial nitric acid plants; consisted of NO, NO_2, N_2O, O_2, N_2, and H_2O; and contained approximately 900–1100 ppm of NO_x, (NO/NO_2 = 2–2.6) 400–600 ppm of N_2O, 2–3 vol.%. of O_2, and 0.3–0.5 vol.% of H_2O. The amount of NH_3 used in the reaction was increased and optimized to the level providing maximum NO_x conversion with minimal NH_3 slip (less than 10 ppm). Thus, the NH_3 concentration was maintained at 0.14–0.15 vol.%, depending on the inlet NO_x concentration.

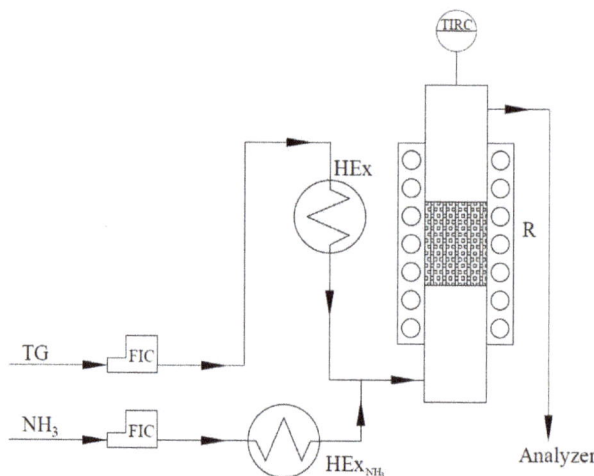

Figure 12. The scheme of the laboratory installation for NH_3-SCR catalytic tests.

In each test, 30 g of the catalyst in a form of pellet (d = 5.0 × 4.8 mm) was placed into the reactor. The activity studies were performed at 250, 300, 350, 400, and 450 °C. The temperature inside the reactor was controlled by a thermocouple installed at the gas outlet from the bed. The research was conducted in GHSV = 4500 and 9000 h^{-1} (tail gas flow 0.15 and 0.3 Nm$^3 \cdot$h^{-1}, respectively). Temperature, GHSV, and shape of catalyst were selected to be as close as possible to conditions prevailing in industrial plants. The measurements of the inlet and outlet concentrations of NO, NO_2, N_2O, and NH_3 were conducted at each temperature after stabilizing the equilibrium conditions and operating parameters. The concentrations of unreacted NO, NO_2, and N_2O were analyzed downstream of the reactor by a GASMET FT-IR analyzer (Vantaa, Finland). NO_x reduction was important in this study; thus, NO and NO_2 concentrations were not considered separately. NO_x conversion was calculated according to Equation (6):

$$X_{NO_x} = \frac{NO_x(in) - NO_x}{NO_x(in)} \cdot 100\% \tag{6}$$

where X_{NO_x}—NO_x conversion, $NO_x(in)$—inlet concentration of NO_x, while NO_x—NO_x concentration in the gas after catalytic reaction.

4. Conclusions

This paper has demonstrated the catalytic potential of protonated or protonated and iron-modified clinoptilolite in the form of pellets in NH_3-SCR within the range of 250–450 °C. Pretreatment with HNO_3 and deposition of iron changed the shape of mesopores and resulted in the formation of secondary porosity. Additionally, deposition of iron caused some interruptions in the order of the zeolite framework. Nevertheless, the crystallinity was not affected by the performed modifications. Catalytic tests were conducted using a gas mixture which reflected industrial conditions. For H-Clin, the maximum conversion of NO_x of over 90% was achieved above 400 °C and GHSV = 4500 h^{-1}. At the load of 9000 h^{-1}, the conversion of NO_x reached more than 60% in the entire temperature range. Satisfactory results obtained for the protonated zeolite without the addition of the active phase can be explained by the natural presence of iron species in the clinoptilolite structure. Regardless of iron loading, NO_x conversion obtained for the catalysts was higher than that of the H-Clin. In the case of Fe-Clin-1 and Fe-Clin-2, NO_x conversion exceeded 90% above 350 °C. Slightly lower NO_x reduction was recorded for Fe-Clin-3. In summary, it was demonstrated that even a single impregnation of natural zeolite (Fe-Clin-1) resulted in the satisfactory catalytic performance, since more than 90% of NO_x conversion was achieved between 350–450 °C. Additionally, it was noted that N_2O concentration decreased by 20% compared to the initial concentration. The strength and significance of our work lies especially in the minimization of the catalyst preparation steps, which is highly beneficial from technological and economical points of view. In summary, it was demonstrated that Fe-clinoptilolite catalysts are advantageous, low-cost, and easy-to-prepare materials that exhibit satisfactory features in the NH_3-SCR process.

Author Contributions: Conceptualization, M.S., M.M. and M.I.; methodology, M.S. and M.I.; resources, K.A.-J.; investigations, M.S. and A.S.-W.; writing—original draft preparation, M.S. and A.S.-W.; writing—review and editing, M.I. and B.S.; supervision, M.I., M.M. and B.S. All authors have read and agreed to the published version of the manuscript.

Funding: M.S. acknowledges the financial support to the program of the Ministry of Science and Higher Education entitled "Implementation Doctorate".

Acknowledgments: A.S.-W. gratefully acknowledges financial support from the National Science Centre Grant Preludium 19 (no. 2020/37/N/ST5/00186) The research results presented by M.M. have been developed with the use of equipment financed from the funds of the "Excellence Initiative-Research University" program at AGH University of Science and Technology. B.S.'s research project was partly supported by the program "Excellence initiative—research university" for the AGH University of Science and Technology.

Conflicts of Interest: The authors declare no conflict of interest. The funders had no role in the design of the study; in the collection, analyses, or interpretation of data; in the writing of the manuscript; or in the decision to publish the results.

References

1. Forzatti, P. Present Status and Perspectives in De-NO_x SCR Catalysis. *Appl. Catal. A Gen.* **2001**, *222*, 221–236. [CrossRef]
2. European Commission. *Integrated Pollution Prevention and Control Reference Document on Best Available Techniques for the Manufacture of Large Volume Inorganic Chemicals—Ammonia, Acids and Fertilisers*; European Commission: Brussels, Belgium, 2007.
3. Kwon, D.W.; Park, K.H.; Ha, H.P.; Hong, S.C. The Role of Molybdenum on the Enhanced Performance and SO_2 Resistance of V/Mo-Ti Catalysts for NH_3-SCR. *Appl. Surf. Sci.* **2019**, *481*, 1167–1177. [CrossRef]
4. Damma, D.; Ettireddy, P.R.; Reddy, B.M.; Smirniotis, P.G. A Review of Low Temperature NH_3-SCR for Removal of NO_x. *Catalysts* **2019**, *9*, 349. [CrossRef]
5. Liu, Q.; Bian, C.; Ming, S.; Guo, L.; Zhang, S.; Pang, L.; Liu, P.; Chen, Z.; Li, T. The Opportunities and Challenges of Iron-Zeolite as NH_3-SCR Catalyst in Purification of Vehicle Exhaust. *Appl. Catal. A Gen.* **2020**, *607*, 117865. [CrossRef]
6. Husnain, N.; Wang, E.; Li, K.; Anwar, M.T.; Mehmood, A.; Gul, M.; Li, D.; Mao, J. Iron Oxide-Based Catalysts for Low-Temperature Selective Catalytic Reduction of NO_x with NH_3. *Rev. Chem. Eng.* **2019**, *35*, 239–264. [CrossRef]

7. Wu, X.; Liu, J.; Liu, X.; Wu, X.; Du, Y. Fabrication of Carbon Doped Cu-Based Oxides as Superior NH_3-SCR Catalysts via Employing Sodium Dodecyl Sulfonate Intercalating CuMgAl-LDH. *J. Catal.* **2022**, *407*, 265–280. [CrossRef]
8. Kobayashi, M.; Kuma, R.; Masaki, S.; Sugishima, N. TiO_2-SiO_2 and V_2O_5/TiO_2-SiO_2 Catalyst: Physico-Chemical Characteristics and Catalytic Behavior in Selective Catalytic Reduction of NO by NH_3. *Appl. Catal. B Environ.* **2005**, *60*, 173–179. [CrossRef]
9. Han, L.; Cai, S.; Gao, M.; Hasegawa, J.Y.; Wang, P.; Zhang, J.; Shi, L.; Zhang, D. Selective Catalytic Reduction of NO_x with NH_3 by Using Novel Catalysts: State of the Art and Future Prospects. *Chem. Rev.* **2019**, *119*, 10916–10976. [CrossRef]
10. Zhu, J.; Liu, Z.; Xu, L.; Ohnishi, T.; Yanaba, Y.; Ogura, M.; Wakihara, T.; Okubo, T. Understanding the High Hydrothermal Stability and NH_3-SCR Activity of the Fast-Synthesized ERI Zeolite. *J. Catal.* **2020**, *391*, 346–356. [CrossRef]
11. Ambrozova, P.; Kynicky, J.; Urubek, T.; Nguyen, V.D. Synthesis and Modification of Clinoptilolite. *Molecules* **2017**, *22*, 1107. [CrossRef]
12. Szymaszek-Wawryca, A.; Díaz, U.; Samojeden, B.; Motak, M. Catalytic Performance of One-Pot Synthesized Fe-MWW Layered Zeolites (MCM-22, MCM-36, and ITQ-2) in Selective Catalytic Reduction of Nitrogen Oxides with Ammonia. *Molecules* **2022**, *27*, 2983. [CrossRef]
13. Szymaszek, A.; Samojeden, B.; Motak, M. Selective Catalytic Reduction of NO_x with Ammonia (NH_3-SCR) over Transition Metal-Based Catalysts-Influence of the Catalysts Support. *Physicochem. Probl. Miner. Process.* **2019**, *55*, 1429–1441. [CrossRef]
14. Dakdareh, A.M.; Falamaki, C.; Ghasemian, N. Hydrothermally Grown Nano-Manganese Oxide on Clinoptilolite for Low-Temperature Propane-Selective Catalytic Reduction of NO_x. *J. Nanoparticle Res.* **2018**, *20*, 309. [CrossRef]
15. Ghasemian, N.; Falamaki, C.; Kalbasi, M.; Khosravi, M. Enhancement of the Catalytic Performance of H-Clinoptilolite in Propane-SCR-NO_x Process through Controlled Dealumination. *Chem. Eng. J.* **2014**, *252*, 112–119. [CrossRef]
16. Ghasemian, N.; Nourmoradi, H. The Study of the Performance of Iranian Clinoptilolite Zeolite in Removal of Nitrogen Oxide (NO_x) from Stack of Industries by Using Selective Catalytic Reduction System (SCR). *J. Ilam Univ. Med. Sci.* **2015**, *23*, 27–35.
17. Ghasemian, N.; Falamaki, C.; Kalbasi, M. Clinoptilolite Zeolite as a Potential Catalyst for Propane-SCR-NO_x: Performance Investigation and Kinetic Analysis. *Chem. Eng. J.* **2014**, *236*, 464–470. [CrossRef]
18. Saramok, M.; Szymaszek, A.; Inger, M.; Antoniak-Jurak, K.; Samojeden, B.; Motak, M. Modified Zeolite Catalyst for a NO_x Selective Catalytic Reduction Process in Nitric Acid Plants. *Catalysts* **2021**, *11*, 450. [CrossRef]
19. Boroń, P.; Chmielarz, L.; Gurgul, J.; Łątka, K.; Gil, B.; Marszałek, B.; Dzwigaj, S. Influence of Iron State and Acidity of Zeolites on the Catalytic Activity of FeHBEA, FeHZSM-5 and FeHMOR in SCR of NO with NH_3 and N_2O Decomposition. *Microporous Mesoporous Mater.* **2015**, *203*, 73–85. [CrossRef]
20. Zhang, T.; Qin, X.; Peng, Y.; Wang, C.; Chang, H.; Chen, J.; Li, J. Effect of Fe Precursors on the Catalytic Activity of Fe/SAPO-34 Catalysts for N_2O Decomposition. *Catal. Commun.* **2019**, *128*, 105706. [CrossRef]
21. Burris, L.E.; Juenger, M.C. The Effect of Acid Treatment on the Reactivity of Natural Zeolites Used as Supplementary Cementitious Materials. *Cem. Concr. Res.* **2016**, *79*, 185–193. [CrossRef]
22. Lin, H.; Liu, Q.L.; Dong, Y.B.; He, Y.H.; Wang, L. Physicochemical Properties and Mechanism Study of Clinoptilolite Modified by NaOH. *Microporous Mesoporous Mater.* **2015**, *218*, 174–179. [CrossRef]
23. Xu, S.; Habib, A.H.; Gee, S.H.; Hong, Y.K.; McHenry, M. Spin Orientation, Structure, Morphology, and Magnetic Properties of Hematite Nanoparticles. *J. Appl. Phys.* **2015**, *117*, 17A315. [CrossRef]
24. Kessouri, A.; Boukoussa, B.; Bengueddach, A.; Hamacha, R. Synthesis of Iron-MFI Zeolite and Its Photocatalytic Application for Hydroxylation of Phenol. *Res. Chem. Intermed.* **2018**, *44*, 2475–2487. [CrossRef]
25. Thommes, M.; Kaneko, K.; Neimark, A.V.; Olivier, J.P.; Rodriguez-Reinoso, F.; Rouquerol, J.; Sing, K.S.W. Physisorption of Gases, with Special Reference to the Evaluation of Surface Area and Pore Size Distribution (IUPAC Technical Report). *Pure Appl. Chem.* **2015**, *87*, 1051–1069. [CrossRef]
26. Vassileva, P.; Voikova, D. Investigation on Natural and Pretreated Bulgarian Clinoptilolite for Ammonium Ions Removal from Aqueous Solutions. *J. Hazard. Mater.* **2009**, *170*, 948–953. [CrossRef]
27. Doula, M.K. Synthesis of a Clinoptilolite-Fe System with High Cu Sorption Capacity. *Chemosphere* **2007**, *67*, 731–740. [CrossRef]
28. Yang, Y.; Gu, Y.; Lin, H.; Jie, B.; Zheng, Z.; Zhang, X. Bicarbonate-Enhanced Iron-Based Prussian Blue Analogs Catalyze the Fenton-like Degradation of p-Nitrophenol. *J. Colloid Interface Sci.* **2022**, *608*, 2884–2895. [CrossRef]
29. Rutkowska, M.; Díaz, U.; Palomares, A.E.; Chmielarz, L. Cu and Fe Modified Derivatives of 2D MWW-Type Zeolites (MCM-22, ITQ-2 and MCM-36) as New Catalysts for $DeNO_x$ Process. *Appl. Catal. B Environ.* **2015**, *168–169*, 531–539. [CrossRef]
30. Korkuna, O.; Leboda, R.; Skubiszewska-Zięba, J.; Vrublevs'ka, T.; Gun'ko, V.M.; Ryczkowski, J. Structural and Physicochemical Properties of Natural Zeolites: Clinoptilolite and Mordenite. *Microporous Mesoporous Mater.* **2006**, *87*, 243–254. [CrossRef]
31. Kumar, A.; Lingfa, P. Sodium Bentonite and Kaolin Clays: Comparative Study on Their FT-IR, XRF, and XRD. *Mater. Today Proc.* **2019**, *22*, 737–742. [CrossRef]
32. Cobzaru, C.; Marinoiu, A.; Cernatescu, C. Sorption of Vitamin C on Acid Modified Clinoptilolite. *Rev. Roum. Chim.* **2015**, *60*, 241–247.
33. Long, R.; Yang, R. Characterization of Fe-ZSM-5 Catalyst for Selective Catalytic Reduction of Nitric Oxide by Ammonia. *J. Catal.* **2000**, *194*, 80–90. [CrossRef]
34. Doula, M.K.; Ioannou, A. The Effect of Electrolyte Anion on Cu Adsorption-Desorption by Clinoptilolite. *Microporous Mesoporous Mater.* **2003**, *58*, 115–130. [CrossRef]

35. Hayati-Ashtiani, M. Use of FTIR Spectroscopy in the Characterization of Natural and Treated Nanostructured Bentonites (Montmorillonites). *Part. Sci. Technol.* **2012**, *30*, 553–564. [CrossRef]
36. Grzybek, J.; Gil, B.; Roth, W.J.; Skoczek, M.; Kowalczyk, A.; Chmielarz, L. Spectrochimica Acta Part A: Molecular and Biomolecular Spectroscopy Characterization of Co and Fe-MCM-56 Catalysts for NH_3-SCR and N_2O Decomposition: An in Situ FTIR Study. *Spectrochim. Acta Part A Mol. Biomol. Spectrosc.* **2018**, *196*, 281–288. [CrossRef]
37. Rivas, F.C.; Rodriguez-Iznaga, I.; Berlier, G.; Ferro, D.T.; Conception-Rosabal, B.; Pertanovskii, V. Fe Speciation in Iron Modified Natural Zeolites as Sustainable Environmental Catalysts. *Catalysts* **2019**, *9*, 866. [CrossRef]

Article

A Promising Cobalt Catalyst for Hydrogen Production

Bogdan Ulejczyk [1,*], Paweł Jóźwik [2], Michał Młotek [1] and Krzysztof Krawczyk [1]

[1] Faculty of Chemistry, Warsaw University of Technology, Noakowskiego 3, 00-664 Warszawa, Poland; michal.mlotek@pw.edu.pl (M.M.); kraw@ch.edu.pl (K.K.)
[2] Faculty of Advanced Technologies and Chemistry, Military University of Technology, Gen. S. Kaliskiego 2, 00-908 Warszawa, Poland; pjozwik@wat.edu.pl
* Correspondence: bulejczyk@ch.pw.edu.pl

Abstract: In this work, a metal cobalt catalyst was synthesized, and its activity in the hydrogen production process was tested. The substrates were water and ethanol. Activity tests were conducted at a temperature range of 350–600 °C, water to ethanol molar ratio of 3 to 5, and a feed flow of 0.4 to 1.2 mol/h. The catalyst had a specific surface area of 1.75 m^2/g. The catalyst was most active at temperatures in the range of 500–600 °C. Under the most favorable conditions, the ethanol conversion was 97%, the hydrogen production efficiency was 4.9 mol (H$_2$)/mol(ethanol), and coke production was very low (16 mg/h). Apart from hydrogen and coke, CO$_2$, CH$_4$, CO, and traces of C$_2$H$_2$ and C$_2$H$_4$ were formed.

Keywords: hydrogen; ethanol; catalyst

1. Introduction

The global consumption of primary energy is increasing rapidly. From 2000 to 2019, primary energy consumption increased from 109,583 TWh to 162,194 TWh [1]. Such an increase in energy consumption causes increased fossil fuel consumption and CO$_2$ emissions, which increased by 11.32 billion tonnes. Renewable energy production is growing all the time, but too slowly to compensate for the increased energy demand. From 2000 to 2019, energy production from renewable sources increased from 2870 to 7017 TWh [1]. During this period, wind and solar energy were developed the most. Nowadays, hydrogen technologies are intensively researched. As hydrogen is not dependent on the weather, it may be a more stable energy source. It is thus necessary to develop effective technology for hydrogen production from the available renewable resources. Ethanol is a readily available raw material obtained from biomass. Ethanol production is perfectly controlled, and is a safe compound. The products of ethanol and water conversion should be hydrogen and carbon dioxide. These are produced in ethanol steam reforming (ESR) and water−gas shift reaction (WGSR). Unfortunately, many competing reactions are also possible. As a result, the post-reaction mixture may contain acetaldehyde, hydrocarbons, carbon monoxide, and coke. Our previous work showed the chemical mechanism in detail [2].

The aim of research on hydrogen production from ethanol and water is to minimize the formation of undesirable products. One of the research paths is the development of a selective and active catalyst. There are many different materials with catalytic properties when producing hydrogen from ethanol. Metals, e.g. cobalt, nickel, copper, and platinum, are often catalysts [3–8]. Of these, cobalt is very attractive because it is cheaper than noble metals. M. Konsolakis et al. [3] reported that cobalt catalysts achieved a higher ethanol conversion than nickel, copper, and iron catalysts. Additionally, cobalt catalysts are characterized by a lower coke production than nickel catalysts [9,10]. Various cobalt catalysts differing in the amount of cobalt, additives, support, or microstructure have been synthesized. Their stability, activity, and selectivity vary depending on the catalyst

Citation: Ulejczyk, B.; Jóźwik, P.; Młotek, M.; Krawczyk, K. A Promising Cobalt Catalyst for Hydrogen Production. *Catalysts* **2022**, *12*, 278. https://doi.org/10.3390/catal12030278

Academic Editors: José Antonio Calles and Angelo Vaccari

Received: 2 February 2022
Accepted: 25 February 2022
Published: 1 March 2022

Publisher's Note: MDPI stays neutral with regard to jurisdictional claims in published maps and institutional affiliations.

Copyright: © 2022 by the authors. Licensee MDPI, Basel, Switzerland. This article is an open access article distributed under the terms and conditions of the Creative Commons Attribution (CC BY) license (https://creativecommons.org/licenses/by/4.0/).

structure. For example, M. Konsolakis et al. [3] reported that the optimal cobalt content is 20%. On the other hand, M. Greluk et al. [11] reported that the conversion of ethanol and the selectivity of ethanol conversion into hydrogen increased with increasing the cobalt content. The highest cobalt content was 30%, and this catalyst was the most active. The reason for these divergent conclusions of the two research groups could be the use of different catalyst support. H. Song et al. [12] compared the properties of cobalt catalysts (10 wt%) supported on three carriers, and reported that the carrier influenced the activity of the cobalt catalyst. Cobalt supported on zirconium oxide was the most active. However, using aluminum oxide as a support caused a lower catalyst activity. The least active was cobalt supported on titanium oxide. The activity of the catalysts was related to the cobalt dispersion (ratio of exposed cobalt atoms to total cobalt atoms). It was the largest on zirconium oxide and the smallest on titanium oxide. On the other hand, the specific surface did not matter, because, for zirconium oxide, it was six times smaller than aluminum oxide and two times smaller than titanium oxide.

Y. Li et al. [13] compared the effect of 16 additives (in the amount of 5%) on alumina-supported cobalt catalysts. The authors reported that the additives influenced the microstructure and activity of the catalyst in different ways. For example, calcium reduced the specific surface area of the catalyst the most, while sodium increased it the most. The size of the cobalt particles was decreased the most by scandium, but titanium increased it the most. These studies did not show a simple relationship between the microstructure and the activity of the catalyst. Na, K, Ni, Cu, Zn, Zr, Ce, La, and Fe increased the ethanol conversion, while Mg, Ca, Ti, Sc, V, Cr, and Mn decreased it. Copper accelerated the coking process the most, while sodium inhibited this process.

It is worth noting that the methodology of the catalyst synthesis is also crucial for the properties of the catalyst. Y. Liu et al. [14] reported that catalyst synthesis methods influenced its activity. The most active catalyst was synthesized by homogeneous precipitation using urea, enabling the highest specific surface area and the smallest cobalt particles. Ch. Wang et al. [15] reported that the addition of various surfactants to the metal precursor solution influenced the structure of the catalyst and its activity. Two surfactants were compared, polyvinylpyrrolidone (PVP) and cetyltrimethyl ammonium bromide (CTAB). Both chemical compounds increased the specific surface area of the catalyst and the porosity, with PVP causing a more significant surface development than CTAB. The specific surface area and porosity increased with increasing surfactant concentration. However, the same correlation between the surfactant concentration and the catalyst activity was not found. The use of CTAB at a lower concentration resulted in the most significant increase in ethanol conversion. On the other hand, for PVP, a greater increase in ethanol conversion was achieved with a lower PVP content. The influence of the catalyst synthesis method on its activity was not only observed for cobalt catalysts. E.V. Matus et al. [16] reported that the method of nickel catalyst synthesis also influenced their activity. Even the change in calcination temperature was significant as it changed the process yield. In addition, in other catalytic processes, the methodology of catalyst synthesis influences its activity, as reported by P. Vacharapong et al. [17].

Generally, despite testing a wide variety of materials, further work is needed to develop an effective and durable ethanol steam reforming catalyst. In this study, a pure metallic cobalt catalyst with no additives was used. The metal catalyst is mechanically resistant, does not crumble, and does not dust. Excellent mechanical properties facilitate all operations with the catalyst, e.g., loading, unloading, and regeneration. The possibility of regeneration is an important factor because the activity of catalysts decreases during operation [3,9,10,13,15–22]. The reason for the decreased activity may be coking [3,10,13,16–21], sintering [3,10,13,15,20], and migration of the active phase [9,22]. Sintered catalysts and catalysts from which the active phase migrated must be replaced. In contrast, coke-coated catalysts can be regenerated. The metal catalyst does not sinter, and metal migration does not change its content on the surface. The metallic cobalt catalyst can be regenerated at high temperatures, which allows the deposited coke to be oxidized rapidly. The metal has an excellent thermal conductivity. Therefore, the temperature in a reactor is easy to control. In reactors with supported catalysts, there are differences between the set-point temperature and the temperature of the catalytic bed. V. Palma et al. [23] reported that these differences depend on the type of carrier and the size of the grains.

In this work, a metal catalyst with a high specific surface area was synthesized. As it was pure cobalt, metal migration did not change its concentration on the catalyst surface. The metal did not sinter, and the melting point of cobalt is much higher than the process temperature. A very good thermal conductivity of cobalt resulted in a very short reactor start-up time. The catalyst bed temperature stabilized within 12 min

2. Results and Discussion
2.1. Catalyst Characterization

The obtained metal catalyst had a specific surface area of 1.75 m^2/g determined by nitrogen adsorption isotherm analysis using the BET isotherm (Brunauer−Emmett−Teller). This method was presented in our earlier articles [2,24]. The catalyst topography was analyzed by scanning electron microscopy (SEM), and its elemental composition was analyzed by energy-dispersive X-ray spectroscopy (EDX). These methods were presented in our earlier article [24]. Figures 1–4 show the catalyst surface and the distribution of elements before and after 20 h of operation. Coke was deposited on the catalyst, as evidenced by the presence of carbon on the catalyst surface after use. Oxygen was also present in the used catalyst. Despite the long-term reduction, there was also oxygen on the surface of the fresh catalyst. However, a comparison of the signal intensities from individual elements shows that oxygen was evenly distributed over the surface of the fresh catalyst. On the contrary, the distribution of elements for the used catalyst indicates that oxygen was present in a greater amount on the surface where carbon was also present. E. L. Viljoen and E. van Steen [25] reported that cobalt is easily oxidized by water. The oxidation process starts at 300 °C. The process conditions for producing hydrogen from ethanol and water enable cobalt oxidation by water. On the other hand, the presence of hydrogen makes it possible to reduce the cobalt oxide. Therefore, it is difficult to conclude whether the correlation between the simultaneous presence of oxygen and carbon results from the facilitated deposition of coke on the cobalt oxide or the difficult access of hydrogen to the surface covered with coke. The carbon structures of the whisker structures, previously reported by G. Słowik et al. [22] and by our group for the plasma-catalytic process [24], were not observed.

(a)

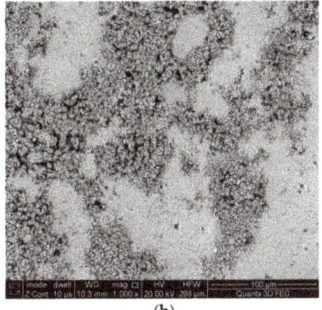
(b)

Figure 1. *Cont.*

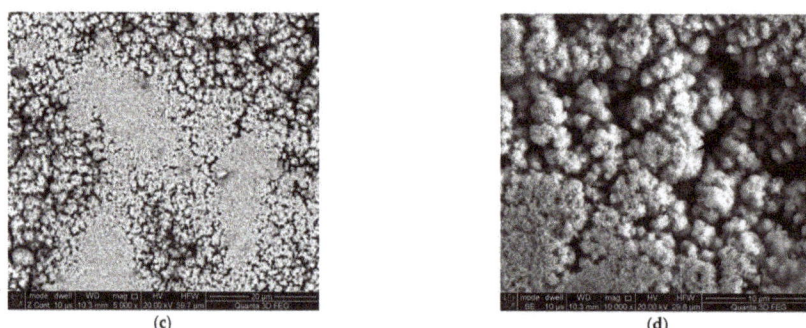

Figure 1. SEM images of the fresh catalyst surface. (**a**) catalyst granule, (**b**) 1000-fold magnification, (**c**) 5000-fold magnification, (**d**) 10,000-fold magnification.

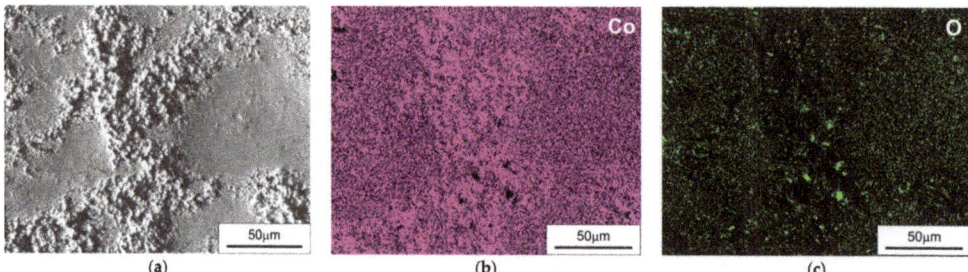

Figure 2. SEM image and EDX mapping analysis of the fresh catalyst. (**a**) SEM picture of the analyzed catalyst's surface, (**b**) distribution of cobalt, and (**c**) distribution of oxygen.

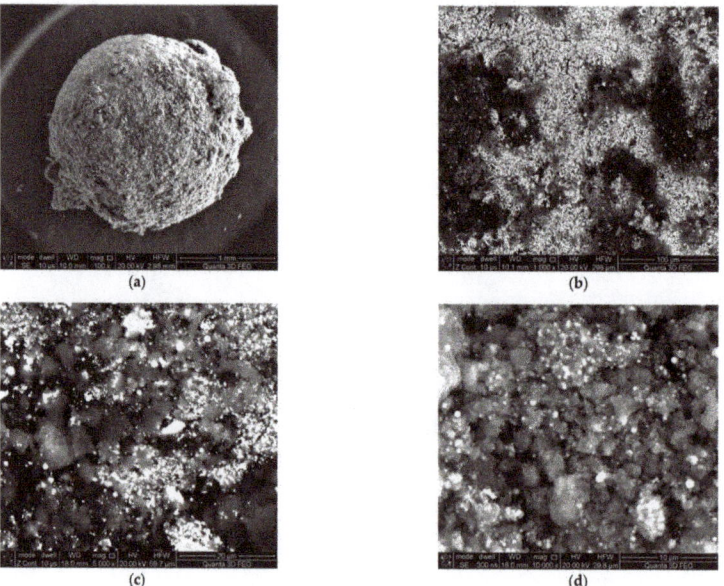

Figure 3. SEM images of the catalyst surface used. (**a**) catalyst granule, (**b**) 1000-fold magnification, (**c**) 5000-fold magnification, (**d**) 10,000-fold magnification.

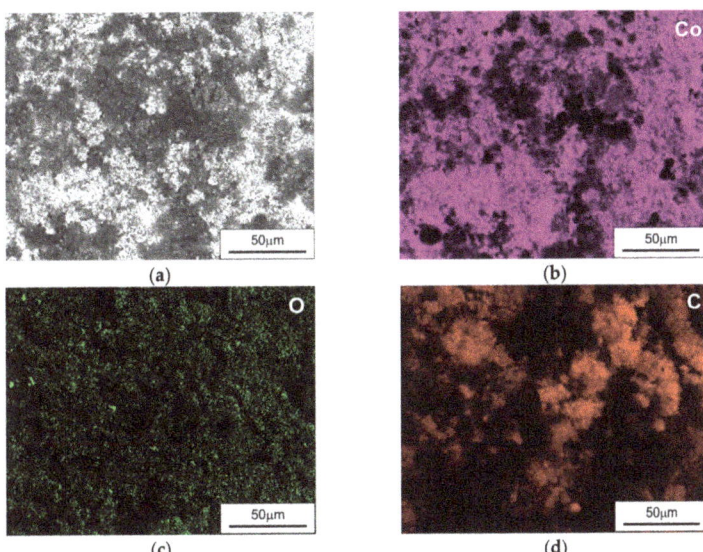

Figure 4. SEM images and EDX mapping analysis of the used catalyst. (**a**) SEM picture of the catalyst surface, (**b**) distribution of deposited cobalt, (**c**) distribution of oxygen, and (**d**) distribution of carbon.

2.2. Gaseous Products

The gaseous products were H_2, CO, CO_2, and CH_4 (Table 1). Moreover, C_2H_2 and C_2H_4 were present in minimal amounts (<0.1%). There was also water vapor (~2%) and ethanol in the cooled gas. The concentration of hydrogen was from 54 to 69%. The concentration of hydrogen increased with the increasing temperature. As the concentration of hydrogen increased, the concentration of CO_2 increased, and the concentrations of CO and CH_4 decreased.

Table 1. The concentration of gaseous products in gases after cooling and coke production.

Feed Flow Rate, mol/h		Temperature, °C	Concentration, %				Coke Production, g/h
Water	Ethanol		H_2	CO	CH_4	CO_2	
0.302	0.099	350	54.23	12.36	14.18	12.11	0.211
		400	57.21	10.64	11.21	13.47	0.260
		450	60.42	7.58	7.69	16.72	0.146
		500	64.24	6.91	4.32	16.8	0.256
		550	66.42	6.37	4.12	17.82	0.267
		600	66.11	7.42	4.01	16.8	0.288
0.601	0.199	350	53.82	14.25	15.34	9.24	0.571
		400	57.81	12.75	11.82	10.24	0.815
		450	60.02	9.51	8.21	14.75	0.418
		500	62.87	8.21	5.89	15.28	0.598
		550	64.38	6.82	5.15	16.2	0.740
		600	64.87	7.02	4.82	15.92	0.769

Table 1. Cont.

Feed Flow Rate, mol/h		Temperature, °C	Concentration, %				Coke Production, g/h
Water	Ethanol		H_2	CO	CH_4	CO_2	
0.901	0.301	350	61.08	14.68	12.05	5.27	1.333
		400	62.47	13.39	10.01	6.85	1.933
		450	64.54	8.52	6.91	12.46	1.560
		500	65.19	7.24	5.82	14.19	1.518
		550	65.42	7.03	5.44	14.28	1.780
		600	65.28	7.52	5.12	14.32	1.541
0.321	0.081	350	61.49	5.82	7.28	18.29	0.075
		400	61.78	5.21	6.92	18.72	0.089
		450	64.1	4.92	5.72	19.24	0.105
		500	67.25	4.42	3.3	19.65	0.145
		550	68.05	4.15	2.95	20.25	0.114
		600	68.01	4.26	3.08	19.98	0.146
0.333	0.066	350	62.49	5.21	6.24	18.61	0.083
		400	62.78	4.81	5.82	19.21	0.051
		450	64.82	4.01	4.53	20.31	0.018
		500	68.08	4.05	3.48	21.05	0.016
		550	68.92	3.82	2.94	21.31	0.016
		600	68.65	3.78	3.15	21.25	0.026

Apart from temperature, an important parameter influencing the concentration of CO, CO_2, and CH_4 was the molar ratio of water to ethanol. As the molar ratio of water to ethanol increased, the concentration of CO_2 increased. With a stoichiometric ratio of water to ethanol of 3, for the reaction (1):

$$C_2H_5OH + 3H_2O \leftrightarrows 6H_2 + 2CO_2 \tag{1}$$

only at a temperature of 450 °C and above was the concentration of CO_2 greater than the concentration of CO and CH_4. Increasing the water to ethanol molar ratio caused the concentration of CO_2, even at 350 °C, to be higher than the concentration of CO and CH_4.

CO and CH_4 are formed in the following reactions (reactions (2) and (3)) [26]:

$$C_2H_5OH + H_2O \leftrightarrows 4H_2 + CO \tag{2}$$

$$C_2H_5OH \leftrightarrows CH_4 + CO + H_2 \tag{3}$$

CO and CH_4 are undesirable products, and water promotes their further conversion [26,27]:

$$CH_4 + H_2O \leftrightarrows CO + 3H_2 \tag{4}$$

$$CO + H_2O \leftrightarrows H_2 + CO_2 \tag{5}$$

The use of water excess increases the production of H_2 and CO_2 and decreases the production of CO and CH_4.

2.3. Coke Formation

Coke production increases rapidly with increasing the feedstock flow (Table 1). The increase in coke production was greater than the increase in substrate flow. This indicates that coke was formed in large amounts and then consumed. The most likely sequence of reactions leading to the formation of coke was initiated by ethanol dehydration (reaction (6)) [26,28]:

$$C_2H_5OH \rightarrow C_2H_4 + H_2O \tag{6}$$

$$C_2H_4 \rightarrow C_2H_2 + H_2 \tag{7}$$

$$C_2H_2 \rightarrow 2C + H_2 \quad (8)$$

when coke was consumed in the hydration reaction (reaction (9)) [26]:

$$C + H_2O \leftrightarrows H_2 + CO \quad (9)$$

The high molar ratio of water to ethanol inhibited the ethanol dehydration reaction and accelerated the coke hydration reaction. Therefore, coke production decreased as the molar ratio of water to ethanol increased.

The coke production reached minimum values at high temperatures, ranging from 450 to 550 °C. Similarly, V. Palma et al. [19] reported that a low temperature promotes coking, whereas B. Banach et al. [29] reported that coke production peaked at temperatures in the range of 560–570 °C, and it depended on the type of carrier. The different results obtained on different catalysts indicate that the properties of the catalyst have a significant influence on the coking process.

2.4. Ethanol Conversion

The metal cobalt catalyst was active from 350 °C. The activity of the catalyst increased rapidly when the temperature increased up to 500 °C (Figures 5 and 6). The further increase in temperature had little effect on the ethanol conversion. Similar changes in the activity of various cobalt catalysts were observed by Y. Li et al. [13].

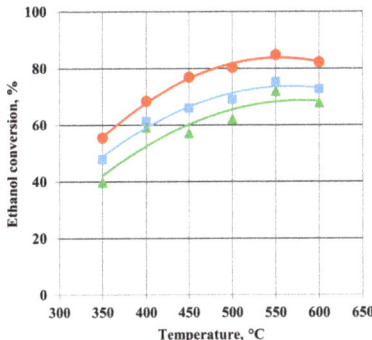

Figure 5. The influence of temperature and feed flow rate on ethanol conversion. Water to ethanol molar ratio in the feed = 3. Feed flow rate: ● = 0.4, ■ = 0.8, and ▲ = 1.2 mol/h.

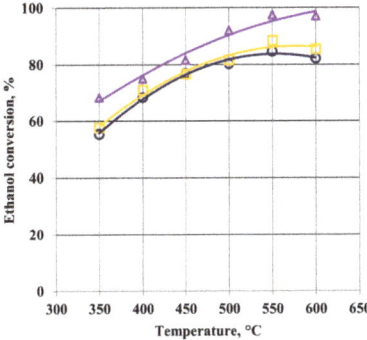

Figure 6. The influence of temperature and feed flow composition on ethanol conversion. Feed flow rate = 0.4 mol/h. Water to ethanol molar ratio in the feed: △ = 5, □ = 4, and ○ = 3.

The ethanol conversion decreased with increasing the feed flow (Figure 5). This resulted from the shortening of the contact time of the reactants with the catalyst. An increase in the molar ratio of water to ethanol increased the ethanol conversion (Figure 6). Ethanol is more expensive than water, so it is worth using the excess water to obtain a higher ethanol conversion. Additional benefits of using excess water are reducing the production of coke, CH_4, and CO and increasing the production of H_2 and CO_2. On the other hand, the disadvantageous effect of using excessive water is the greater energy consumption necessary to heat and evaporate the water. Due to the energy cost in the industrial process of producing hydrogen from natural gas, the H_2O/C molar ratio is 2.5 [27], and efforts have been made to reduce it in new installations [30]. Q. Shen et al. [31] reported that the highest ethanol conversion was achieved with a water to ethanol molar ratio of 5 (H_2O/C molar ratio 2.5). The use of more water did not increase the ethanol conversion.

2.5. Hydrogen Yield

Figures 7 and 8 show the effect of temperature, feed flow rate, and feed stream composition on the hydrogen yield. The hydrogen yield increased with increasing the temperature, but already at 500 °C, it reached a high value. The increase in temperature to 600 °C had little effect on the hydrogen yield. The hydrogen yield decreased with increasing the feed flow rate. In contrast, the increase in the molar ratio of water to ethanol increased the hydrogen yield. The highest hydrogen production efficiency was 4.9 for high temperatures, the lowest flow, and the highest molar ratio of water to ethanol. Q. Shen et al. [31] also reported that the highest hydrogen yield was achieved at a high temperature and a high molar ratio of water to ethanol. The highest hydrogen yield was ~4.8 and was achieved for a higher molar ratio of water to ethanol compared to that used in this work.

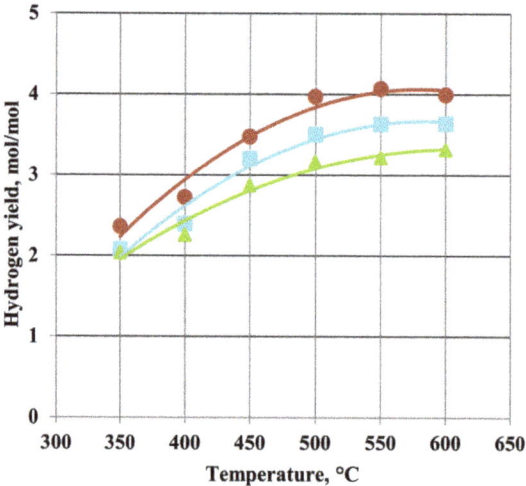

Figure 7. The influence of temperature and feed flow rate on hydrogen yield. Water to ethanol molar ratio in the feed = 3. Feed flow rate: ● = 0.4, ▪ = 0.8, and ▲ = 1.2 mol/h.

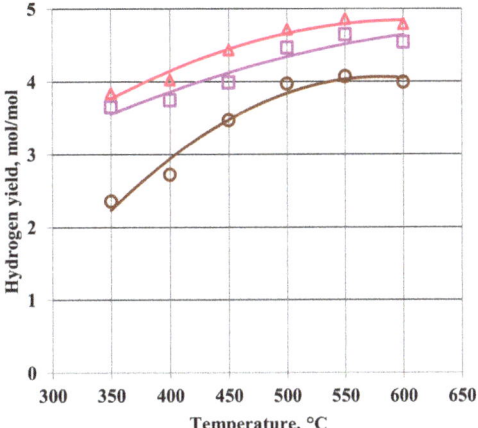

Figure 8. The influence of temperature and feed flow composition on the hydrogen yield. Feed flow rate = 0.4 mol/h. Water to ethanol molar ratio in the feed: △ = 5, ☐ = 4, and o = 3.

3. Materials and Methods

3.1. Catalyst Preparation

The catalyst preparation scheme is shown in Figure 9. Firstly, powdered metallic cobalt was dissolved in nitric acid.

$$Co + 2HNO_3 \rightarrow Co(NO_3)_2 + H_2 \tag{10}$$

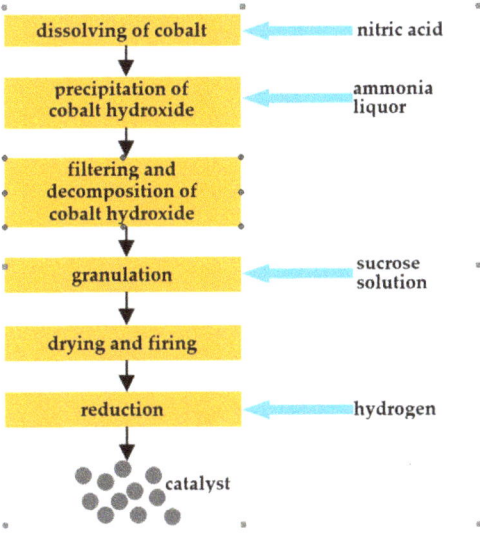

Figure 9. Schematic illustration of the cobalt catalyst preparation procedure.

Ammonia liquor was added to the obtained cobalt nitrate, and the cobalt hydroxide precipitated.

$$Co(NO_3)_2 + 2H_2O + 2NH_3 \rightarrow Co(OH)_2 + 2NH_4NO_3 \tag{11}$$

The obtained cobalt hydroxide was filtered and calcined at 600 °C for 5 h.

$$6Co(OH)_2 + O_2 \rightarrow 2Co_3O_4 + 6H_2O \qquad (12)$$

The obtained cobalt oxide was wetly granulated in a disk granulator using a 20% sucrose solution as a granulating liquid. Next, the 0.8–2 mm fraction was separated on the sieves. The granules were dried for 24 h. Then, the granules were fired at 1250 °C for 5 h. After cooling down the granules, the 0.8–2 mm fraction was separated and reduced with hydrogen at a temperature of 300 °C for 6 h.

$$Co_3O_4 + 4H_2 \rightarrow 3Co + 4H_2O \qquad (13)$$

3.2. Catalyst Activity Tests

The installation for conducting the catalyst activity tests is shown in Figure 10. A catalyst sample (2 g) was placed in a tubular quartz reactor with an internal diameter of 11 mm. After the substrates were introduced into the reactor, the heating started. The catalyst could not be preheated in the air as cobalt oxidizes at 80 °C. Catalyst activity tests were carried out at a temperature range of 350–600 °C. The temperature of 350 °C was the lowest temperature at which the catalyst showed activity. The maximum test temperature was set at 600 °C, as the energy cost of hydrogen production increased with increasing the temperature. Currently, research is being carried out on new solid oxide fuel cells (SOFCs) that may operate at temperatures up to 600 °C [32,33].

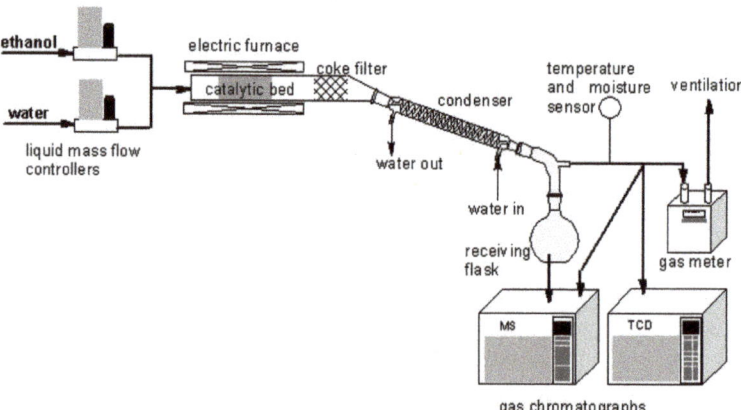

Figure 10. Scheme of the installation for the catalyst activity tests.

The molar ratio of water to ethanol ranged from 3 to 5. The molar feed flow rate ranged from 0.4 to 1.2 mol/h. The methods and apparatus used in the catalyst activity tests were described in detail in our previous works [2,24,34,35].

The ethanol conversion (X, %) and the hydrogen yield (Y, mol(H_2)/mol(C_2H_5OH)) were calculated from Formulas ((14) and (15)):

$$X = (F_{in}[C_2H_5OH] - F_{out}[C_2H_5OH])/F_{in}[C_2H_5OH]*100\%, \qquad (14)$$

$$Y = F_{out}[H_2]/(F_{in}[C_2H_5OH] - F_{out}[C_2H_5OH]), \qquad (15)$$

$F_{in}[C_2H_5OH]$: ethanol feed flow, mol/h,
$F_{out}[C_2H_5OH]$: flow of the ethanol leaving the reactor, mol/h,
$F_{out}[H_2]$: hydrogen production rate, mol/h.

4. Conclusions

The cobalt catalyst is a promising catalyst for hydrogen production from a water and ethanol mixture. The metal catalyst is resistant to sintering and active phase migration. It enables high ethanol conversion and high hydrogen production efficiency. The conditions conducive to obtaining high conversion and yield values are high temperature, low feed flow, and excess of water. The highest ethanol conversion and hydrogen production efficiency were achieved for a temperature of 550 °C, substrate flow of 0.4 mol/h, and water to ethanol molar ratio of 5. Under these conditions, the coke production was low, at 16 mg/h. The use of water in excess was beneficial, as increasing the water to ethanol molar ratio increased the concentration of H_2 and CO_2. In contrast, the concentration of CH_4 and CO decreased. The high activity of the cobalt catalyst at a temperature of 500–600 °C means that it can be used in the production of hydrogen to supply solid oxide fuel cells operating at low temperatures.

Author Contributions: Conceptualization, B.U.; methodology, B.U., M.M. and P.J.; validation, B.U., K.K., P.J. and M.M.; formal analysis, B.U.; investigation, B.U. and P.J.; resources, B.U. and P.J.; data curation, B.U.; writing—original draft preparation, B.U.; writing—review and editing, B.U.; supervision, B.U.; funding acquisition, K.K. All authors have read and agreed to the published version of the manuscript.

Funding: This research was funded by the Warsaw University of Technology.

Data Availability Statement: Not applicable.

Conflicts of Interest: The authors declare no conflict of interest.

References

1. Ritchie, H.; Roser, M. Energy, Published Online at OurWorldInData.org. Available online: https://ourworldindata.org/energy (accessed on 1 February 2022).
2. Ulejczyk, B.; Nogal, Ł.; Młotek, M.; Falkowski, P.; Krawczyk, K. Hydrogen production from ethanol using a special multi-segment plasma-catalytic reactor. *J. Energy Inst.* **2021**, *95*, 179–186. [CrossRef]
3. Konsolakis, M.; Ioakimidis, Z.; Kraia, T.; Marnellos, G.E. Hydrogen Production by Ethanol Steam Reforming (ESR) over CeO_2 Supported Transition Metal (Fe, Co, Ni, Cu) Catalysts: Insight into the Structure-Activity Relationship. *Catalysts* **2016**, *6*, 39. [CrossRef]
4. Matus, E.; Sukhova, O.; Ismagilov, I.; Kerzhentsev, M.; Stonkus, O.; Ismagilov, Z. Hydrogen Production through Autothermal Reforming of Ethanol: Enhancement of Ni Catalyst Performance via Promotion. *Energies* **2021**, *14*, 5176. [CrossRef]
5. Rajabi, Z.; Jones, L.; Martinelli, M.; Qian, D.; Cronauer, D.C.; Kropf, A.J.; Watson, C.D.; Jacobs, G. Influence of Cs Promoter on Ethanol Steam-Reforming Selectivity of Pt/m-ZrO_2 Catalysts at Low Temperature. *Catalysts* **2021**, *11*, 1104. [CrossRef]
6. Hamryszak, Ł.; Kulawska, M.; Madej-Lachowska, M.; Śliwa, M.; Samson, K.; Ruggiero-Mikołajczyk, M. Copper Tricomponent Catalysts Application for Hydrogen Production from Ethanol. *Catalysts* **2021**, *11*, 575. [CrossRef]
7. Cifuentes, B.; Valero, M.F.; Conesa, J.A.; Cobo, M. Hydrogen Production by Steam Reforming of Ethanol on Rh-Pt Catalysts: Influence of CeO_2, ZrO_2, and La_2O_3 as Supports. *Catalysts* **2015**, *5*, 1872–1896. [CrossRef]
8. da Costa-Serra, J.F.; Navarro, M.T.; Rey, F.; Chica, A. Sustainable Production of Hydrogen by Steam Reforming of Ethanol Using Cobalt Supported on Nanoporous Zeolitic Material. *Nanomaterials* **2020**, *10*, 1934. [CrossRef]
9. Greluk, M.; Rotko, M.; Turczyniak-Surdacka, S. Enhanced catalytic performance of La_2O_3 promoted Co/CeO_2 and Ni/CeO_2 catalysts for effective hydrogen production by ethanol steam reforming. *Renew. Energy* **2020**, *155*, 378–395. [CrossRef]
10. Contreras, J.L.; Figueroa, A.; Zeifert, B.; Salmones, J.; Fuentes, G.A.; Vázquez, T.; Angeles, D.; Nuño, L. Production of hydrogen by ethanol steam reforming using Ni–Co-ex-hydrotalcite catalysts stabilized with tungsten oxides. *Int. J. Hydrogen Energy* **2021**, *46*, 6474–6493. [CrossRef]
11. Greluk, M.; Rotko, M.; Słowik, G.; Turczyniak-Surdacka, S. Hydrogen production by steam reforming of ethanol over Co/CeO_2 catalysts: Effect of cobalt content. *J. Energy Inst.* **2019**, *92*, 222–238. [CrossRef]
12. Song, H.; Zhang, L.; Watson, R.B.; Braden, D.; Ozkan, U.S. Investigation of bio-ethanol steam reforming over cobalt-based catalysts. *Catal. Today* **2007**, *129*, 346–354. [CrossRef]
13. Li, Y.; Zhang, Z.; Jia, P.; Dong, D.; Wang, Y.; Hu, S.; Xiang, J.; Liu, Q.; Hu, X. Ethanol steam reforming over cobalt catalysts: Effect of a range of additives on the catalytic behaviors. *J. Energy Inst.* **2020**, *93*, 165–184. [CrossRef]
14. Liu, Y.; Murata, K.; Inaba, M. Steam Reforming of Bio-Ethanol to Produce Hydrogen over Co/CeO_2 Catalysts Derived from $Ce_{1-x}Co_xO_{2-y}$ Precursors. *Catalysts* **2016**, *6*, 26. [CrossRef]

15. Wang, C.; Wang, Y.; Chen, M.; Hu, J.; Yang, Z.; Zhang, H.; Wang, J.; Liu, S. Hydrogen production from ethanol steam reforming over Co–Ce/sepiolite catalysts prepared by a surfactant assisted coprecipitation method. *Int. J. Hydrogen Energy* **2019**, *44*, 26888–26904. [CrossRef]
16. Matus, E.V.; Okhlopkova, L.B.; Sukhova, O.B.; Ismagilov, I.Z.; Kerzhentsev, M.A.; Ismagilov, Z.R. Effects of preparation mode and doping on the genesis and properties of Ni/Ce$_{1-x}$M$_x$O$_y$ nanocrystallites (M = Gd, La, Mg) for catalytic applications. *J. F Nanopart. Res.* **2019**, *21*, 11. [CrossRef]
17. Vacharapong, P.; Arayawate, S.; Katanyutanon, S.; Toochinda, P.; Lawtrakul, L.; Charojrochkul, S. Enhancement of Ni Catalyst Using CeO$_2$–Al$_2$O$_3$ Support Prepared with Magnetic Inducement for ESR. *Catalysts* **2020**, *10*, 1357. [CrossRef]
18. Chen, Y.-J.; Huang, S.-H.; Uan, J.-Y.; Lin, H.-T. Synthesis of Catalytic Ni/Cu Nanoparticles from Simulated Wastewater on Li–Al Mixed Metal Oxides for a Two-Stage Catalytic Process in Ethanol Steam Reforming: Catalytic Performance and Coke Properties. *Catalysts* **2021**, *11*, 1124. [CrossRef]
19. Palma, V.; Ruocco, C.; Meloni, E.; Ricca, A. Influence of Catalytic Formulation and Operative Conditions on Coke Deposition over CeO$_2$-SiO$_2$ Based Catalysts for Ethanol Reforming. *Energies* **2017**, *10*, 1030. [CrossRef]
20. Mhadmhan, S.; Natewong, P.; Prasongthum, N.; Samart, C.; Reubroycharoen, P. Investigation of Ni/SiO$_2$ Fiber Catalysts Prepared by Different Methods on Hydrogen production from Ethanol Steam Reforming. *Catalysts* **2018**, *8*, 319. [CrossRef]
21. Palma, V.; Ruocco, C.; Meloni, E.; Gallucci, F.; Ricca, A. Enhancing Pt-Ni/CeO$_2$ performances for ethanol reforming by catalyst supporting on high surface silica. *Catal. Today* **2018**, *307*, 175–188. [CrossRef]
22. Słowik, G.; Greluk, N.; Rotko, M.; Machocki, A. Evolution of the structure of unpromoted and potassium-promoted ceriasupported nickel catalysts in the steam reforming of ethanol. *Appl. Catal. B Environ.* **2018**, *221*, 490–509. [CrossRef]
23. Palma, V.; Ruocco, C.; Castaldo, F.; Ricca, A.; Boettge, D. Ethanol steam reforming over bimetallic coated ceramic foams: Effect of reactor configuration and catalytic support. *Int. J. Hydrogen Energy* **2015**, *37*, 12650–12662. [CrossRef]
24. Ulejczyk, B.; Nogal, Ł.; Jóźwik, P.; Młotek, M.; Krawczyk, K. Plasma-Catalytic Process of Hydrogen Production from Mixture of Methanol and Water. *Catalysts* **2021**, *11*, 864. [CrossRef]
25. Viljoen, E.L.; van Steen, E. Rate of Oxidation of a Cobalt Catalyst in Water and Water/Hydrogen Mixtures: Influence of Platinum as a Reduction Promoter. *Catal. Lett.* **2009**, *133*, 8–13. [CrossRef]
26. de Souza, G.; Balzaretti, N.M.; Marcílio, N.R.; Perez-Lopez, O.W. Decomposition of Ethanol Over Ni-Al Catalysts: Effect of Copper Addition. *Procedia Eng.* **2012**, *42*, 335–345. [CrossRef]
27. Rostrup-Nielsen, J.R. Catalytic Steam Reforming. In *Catalysis Science and Technology*; Anderson, J.R., Boudart, M., Eds.; Springer-Verlag: Berlin/Heidelberg, Germany, 1984; Volume 5, pp. 3–117. [CrossRef]
28. Rye, R.R.; Hansen, R.S. Flash Decomposition of Acetylene, Ethylene, Ethane, and Methane on Tungsten. *J. Chem. Phys.* **1969**, *50*, 3585–3595. [CrossRef]
29. Banach, B.; Machocki, A.; Rybak, P.; Denis, A.; Grzegorczyk, W.; Gac, W. Selective production of hydrogen by steam reforming of bio-ethanol. *Catal. Today* **2011**, *176*, 28–35. [CrossRef]
30. Eyalarasan, K.; Tesfamariam, M.D.; Meleake, H.; Gebreyonas, A. Design of Process Plant for Producing Hydrogen from Steam Reforming of Natural Gas. *Int. J. Eng. Res. Technol.* **2013**, *2*, 746–754.
31. Shen, Q.; Jiang, Y.; Li, S.; Yang, G.; Zhang, H.; Zhang, Z.; Pan, X. Hydrogen production by ethanol steam reforming over Ni-doped LaNi$_x$Co$_{1-x}$O$_{3-\delta}$ perovskites prepared by EDTA-citric acid sol–gel method. *J. Sol-Gel Sci. Technol.* **2021**, *99*, 420–429. [CrossRef]
32. Thieu, C.A.; Yang, S.; Ji, H.I.; Kim, H.; Yoon, K.J.; Lee, J.H.; Son, J.W. Effect of secondary metal catalysts on butane internal steam reforming operation of thin-film solid oxide fuel cells at 500–600 °C. *Appl. Catal. B Environ.* **2020**, *263*, 118349. [CrossRef]
33. Fallah Vostakola, M.; Amini Horri, B. Progress in Material Development for Low-Temperature Solid Oxide Fuel Cells: A Review. *Energies* **2021**, *14*, 1280. [CrossRef]
34. Ulejczyk, B.; Nogal, Ł.; Młotek, M.; Krawczyk, K. Hydrogen production from ethanol using dielectric barrier discharge. *Energy* **2019**, *174*, 261–268. [CrossRef]
35. Ulejczyk, B.; Nogal, Ł.; Młotek, M.; Krawczyk, K. Enhanced production of hydrogen from methanol using spark discharge generated in a small portable reactor. *Energy Rep.* **2022**, *8*, 183–191. [CrossRef]

Article

Effect of Potassium Promoter on the Performance of Nickel-Based Catalysts Supported on MnO$_x$ in Steam Reforming of Ethanol

Magdalena Greluk [1,*], Marek Rotko [1], Grzegorz Słowik [1], Sylwia Turczyniak-Surdacka [2], Gabriela Grzybek [3], Kinga Góra-Marek [3] and Andrzej Kotarba [3]

1. Department of Chemical Technology, Faculty of Chemistry, Maria Curie-Sklodowska University in Lublin, Maria Curie-Sklodowska Sq. 3, 20-031 Lublin, Poland; marek.rotko@mail.umcs.pl (M.R.); grzegorz.slowik@mail.umcs.pl (G.S.)
2. Biological and Chemical Research Centre, University of Warsaw, 101 Żwirki i Wigury Street, 20-089 Warsaw, Poland; sturczyniak@cnbc.uw.edu.pl
3. Faculty of Chemistry, Jagiellonian University, Gronostajowa 2, 30-387 Krakow, Poland; g.grzybek@uj.edu.pl (G.G.); kinga.gora-marek@uj.edu.pl (K.G.-M.); ak@uj.edu.pl (A.K.)
* Correspondence: magdalena.greluk@mail.umcs.pl; Tel.: +48-81-537-55-14; Fax: +48-81-537-55-65

Abstract: The effect of a potassium promoter on the stability of and resistance to a carbon deposit formation on the Ni/MnO$_x$ catalyst under SRE conditions was studied at 420 °C for different H$_2$O/EtOH molar ratios in the range from 4/1 to 12/1. The catalysts were prepared by the impregnation method and characterized using several techniques to study their textural, structural, and redox properties before being tested in a SRE reaction. The catalytic tests indicated that the addition of a low amount of potassium (1.6 wt.%) allows a catalyst with high stability to be obtained, which was ascribed to high resistance to carbon formation. The restriction of the amount of carbon deposits originates from the potassium presence on the Ni surface, which leads to (i) a decrease in the number of active sites available for methane decomposition and (ii) an increase in the rate of the steam gasification of carbon formed during SRE reactions.

Keywords: hydrogen production; ethanol steam reforming; nickel-based catalyst; manganese oxides; potassium promoter

1. Introduction

Hydrogen does not affect the environment adversely, unlike fossil fuels and nuclear power; thus, it is considered a clean next-generation energy source with great potential for commercialization. Unfortunately, it is not freely available in nature and it must be produced. For hydrogen to become a truly sustainable energy source its production from renewable resources should be promoted. Ethanol is an attractive choice of raw material for producing hydrogen because, besides its low ecotoxicity, it can be produced from biomass and utilized without the net addition of carbon dioxide to the atmosphere. Ethanol produced by biomass fermentation has a high ratio of water to ethanol, which is suitable for the steam reforming reaction. In the case of this process, it is not necessary to carry out the distillation of fermented to obtain pure ethanol. Therefore, it is possible to directly use bioethanol obtained from biomass fermentation in steam reforming reactions, which can save lots of money and energy for purification [1–4].

One of the greatest challenges in the production of renewable hydrogen from bioethanol using the catalytic steam reforming of ethanol (SRE) process is the development of a catalyst with the desirable activity, selectivity, and stability. Transition metals have widespread applications as heterogeneous catalysts for a number of reactions and two of them, i.e., cobalt and nickel, possess an excellent performance in the SRE reaction due to their high C–C bond cleavage activity. The interaction of a transition metal with a support has a significant influence on the structure definition and the catalytic activity in the SRE process. Supporting metals might allow the production very active nanoparticles stabilized towards sintering and somehow activated by the interaction with the support. The most often used supports for cobalt- or nickel-containing catalysts in the SRE process are oxides [5–8]. Alumina is one of the most widely used supports in catalysis due to its high surface area, which allows a great dispersion of the active phase [3], and its application as the support of cobalt- or nickel-based catalysts in the SRE process has been extensively studied [9–14]. However, the acidic character of alumina may promote ethanol dehydration to ethylene, which can lead to carbon formation on a catalyst surface. On the other hand, redox supports like ceria and ceria-containing mixed oxides have been proposed to prevent the carbon deposits on the catalytic structure under SRE conditions, due to their high oxygen storage capacity [7,15]. Although the previously mentioned redox support oxides are highly favorable for ethanol reforming, the unique redox property and strong oxygen storage and release ability of manganese oxides (MnO_x) were practically not investigated in this process. It is all the more remarkable that manganese oxide-based systems represent a very intriguing and peculiar class of support with both high chemical stability and the enhanced redox behavior, allowing the stabilization of various intermediate manganese oxidation states (MnO_2, Mn_5O_8, Mn_2O_3, Mn_3O_4, and MnO) and crystalline phases, which provides potential suitability for catalytic reactions [8,16,17].

In the literature, there are only a few studies on the application of manganese oxides in an SRE reaction over cobalt- or nickel-based catalysts. Most of these reports concern the application of this oxide as a modifier of cobalt-based catalysts [5,18,19]. Lee et al. [18] showed that the introduction of manganese oxides to the $Co_{10}Si_{90}$MCM-48 material had a favorable effect on its catalytic performance in the SRE process by preventing the catalytic deactivation caused by the sintering of cobalt particles. Moreover, according to the same authors [18], manganese oxides appeared to provide oxygen to cobalt species, resulting in increases in hydrogen production and the suppression of carbon monoxide production. The same conclusions were drawn by Kwak et al. [19] for the role of manganese oxide in the spinel structured Co_2MnO_4 of Co_2MnO_4/SBA-15 material during its catalytic performance in the SRE reaction. However, not only positive effects on the performance of manganese-oxide-modified cobalt-based catalysts under the SRE conditions were observed. As was presented by Li et al. [5], an addition of manganese oxides to the Co/Al_2O_3 catalyst suppressed its activity in the SRE process. Furthermore, a few reports concerning the application of manganese oxide as a support of cobalt- and nickel-based catalysts can be found in the literature [2,20,21]. Both cobalt- and nickel-based catalysts supported on two structured manganese oxides, i.e., birnessite and todorokite, were found to be highly active and selective to produce hydrogen in the SRE reaction [2,20]. In addition, the cryptomelane-based cobalt-manganese catalyst showed high ethanol conversion and selectivity to hydrogen and carbon dioxide in the SRE reaction in the temperature range of 390–480 °C. However, it was observed only at the initial stage of the process. With the increase in the time-on-stream, its activity decreased with a simultaneous increase in the acetaldehyde production, which was ascribed to deactivation due to carbon formation [21]. Sohrabi and Irankhah [22] showed that $Ni/CeMnO_2$ catalyst was active in the SRE process and its activity was improved by the addition of copper or iron promoters.

Carbon deposition is the major deactivation mechanism of the ethanol steam reforming catalysts. Although reducibility and oxygen transfer capacity of support have shown to be fundamental in the keeping the active phase surface free of carbon deposits [3], the addition of alkali is another solution that is also commonly used to reduce the problem of carbon formation under the SRE conditions [22–33]. In general, the application of alkali promoters allows a decrease in the rate of formation of carbon deposits on the surface of the catalysts during the SRE process [25,31–34]. However, Grzybek et al. [29] indicated that the rate of carbon depositions on the surface of a potassium-doped CoZn | a-Al_2O_3 catalyst is even higher compared to the undoped sample and its enhanced stability under SRE conditions results from a stronger interaction between the cobalt nanocrystals and the alumina support in the presence of a potassium promoter, stabilizing the active phase against the driven detachment and its subsequent encapsulation by the growing carbon deposit. Moreover, Słowik et al. [28] showed that the presence of a potassium promoter does not protect the Ni/CeO$_2$ catalyst against carbon deposit formation and even worsens its stability under SRE conditions compared to the unpromoted sample. According to the authors, the addition of a potassium promoter influences the structure and degree of graphitization of the formed carbon deposits, leading to its faster deactivation.

Considering the above points, the purpose of the present work was to study the performance of the Ni/MnO$_x$ and KNi/MnO$_x$ catalysts under an SRE reaction and to determine the role of a potassium promoter on the stability and resistance to carbon deposit formation. The catalysts were prepared by the impregnation method and characterized using several techniques in order to study their textural, structural, and redox properties before being tested in an SRE reaction.

2. Experimental Procedure

2.1. Catalyst Synthesis

(CH$_3$COO)$_2$Mn·4 H$_2$O ($\geq$97.0%, Sigma-Aldrich, Darmstadt, Germany), Ni(NO$_3$)$_2$·6 H$_2$O (>97.0%, Sigma-Aldrich, Darmstadt, Germany), KNO$_3$ ($\geq$99.0%, Merck, Darmstadt, Germany), and (NH$_4$)$_2$CO$_3$ (100%, Avantor, Gliwice, Poland) were used without further purification.

In the first step, the MnO$_x$ support was prepared using the conventional precipitation method from a 0.5 mol/L manganese acetate aqueous solution. Precipitation was accomplished at 40 °C with the addition of 1 mol/L ammonium carbonate solution, drop by drop, up to a pH of 8, under continuous stirring in suspension. After the aging of the precipitate at 60 °C for 2 h, the suspension was filtrated. The filtrate was washed with absolute ethanol in order to remove water from the precipitates. The obtained solid was dried overnight at 110 °C and then calcined at 500 °C in air.

In the second step, the Ni/MnO$_x$ catalysts were prepared using the wet impregnation method, with 10 g of a finely grinded MnO$_x$ solid and solutions containing 5.505 g of nickel nitrate. The mixture was dried at 110 °C for 12 h and then calcined at 500 °C in air to decompose the nickel nitrate into the nickel oxide precursor. The loading of cobalt or nickel to MnO$_x$ was 10 wt%.

In the final step, the KNi/MnO$_x$ catalysts were prepared using the wet impregnation method, with 5 g of a finely grinded Ni/MnO$_x$ solid and solutions containing 0.264 g of potassium nitrate. The mixture was dried at 110 °C for 12 h and then calcined at 500 °C in air to decompose the potassium nitrate. The loading of potassium to Ni/MnO$_x$ was 2 wt%.

2.2. Catalysts Characterization

The chemical composition of the cobalt- and nickel-based catalysts was determined by the X-ray fluorescence method, using an Axios mAX (PANalytical, Malvern, UK) fluorescence spectrometer. The textural properties of the support and catalysts were determined from nitrogen adsorption-desorption isotherms at −196 °C using ASAP 2405N (Micromeritics, Norcross, GA, USA). The specific area was obtained using the BET method and pore volume and diameter were obtained using the BJH method. The linear region in the BET plots is between p/p^0 = 0 and 0.25. Prior to the analysis, the samples were outgassed at 200 °C. The X-ray diffraction (XRD) patterns were collected with an Empyrean X-ray (PANalytical, Malvern, UK) diffractometer with Cu Kα radiation (λ = 0.154 nm) in the 20–100° 2θ range and a step of 0.026°. In order to analyze samples in their reduced form, prior to the analysis, they were activated in situ at 500 °C with hydrogen in an XRK 900 reactor chamber (Anton Paar, St. Albans, UK). The morphology and lattice profiles of all the catalysts after their reduction were characterized by an electron transmission microscope Titan G2 60–300 kV (FEI Company, Hillsboro, OR, USA). The fast Fourier transform (FFT) was obtained to determine which d-spacing corresponded to which crystalline species. The mapping was carried out in the STEM mode by collecting a point-by-point EDS spectrum of each of the corresponding pixels on the map. By measuring the size of at least 200 particles from the HRTEM images, particle size distributions and mean particle size data were obtained. Prior to the analysis, the catalysts were reduced in a fixed-bed reactor with hydrogen at 500 °C and transferred in a closed reactor to a glove box filled with argon, which prevented the catalyst from oxidation. Then, a TEM vacuum transfer holder was used to transfer the catalysts laid on a copper grid in the glove box, from the aforementioned preparation station into a TEM microscope in argon atmosphere. SEM images of the catalysts used in the SRE process were taken using a FIB-SEM Crossbeam 540 FEG (Zeiss, Jena, Germany) with an acceleration voltage of 1.8 kV. The samples were prepared by drying the solution of catalyst diluted in ethanol dropped onto carbon tape. TEM images of catalysts used in the SRE process were obtained in an electron transmission microscope, Titan G2 60–300 kV (FEI Company). Samples were suspended in ethanol and exposed to ultrasonic vibrations to decrease aggregation. A drop of the resultant mixture was laid on the copper grid covered with lacey formvar and stabilized with carbon.

H_2-TPR measurements were carried out with an AutoChem II 2920 (Micromeritics, Norcross, GA, USA) analyzer. Typically, a 50 mg sample was placed in a quartz U-tube reactor and heated to 750 °C at 10 °C/min under the mixed reduction gas flow (5% H_2/He, 30 mL·min^{-1}). The hydrogen consumption was monitored by a thermal conductivity detector (TCD).

The acidity of the samples was determined by the FT-IR pyridine adsorption studies. The sample was pressed into the form of a disc, then placed in a custom-made quartz IR cell and in situ evacuated, then activated at 500 °C in a pure hydrogen stream for 1 h.

To estimate the impact of potassium on the catalyst's electron-donor properties, the work function was determined. The contact potential difference (VCPD) was measured by means of the dynamic condenser method of Kelvin using a KP6500 probe (McAllister Technical Services, Berkeley, CA, USA). A stainless-steel plate was used as a reference electrode (ϕ_{ref} = 4.3 eV) at 3 mm in diameter. The measurement parameters were: the vibration frequency at 120 Hz and the amplitude at 40 a.u. The measurements were performed at ambient conditions (room temperature, atmospheric pressure). The work function values were calculated based on the equation: $V_{CPD} = \phi_{ref} - \phi_{sample}$.

X-ray photoelectron spectroscopy (XPS) studies were performed in a Kratos Axis Supra Spectrometer equipped with a monochromatized Al source operating at 150 W over the samples mounted on conductive double-side Cu tape. The pass energy of the analyzer was set to 160 eV (energy step 1.0 eV) for the survey scan and 20 eV (energy step 0.1 eV) for high-resolution Ni 2p, Mn 2p, O 1s, C 1s, and K 2p spectra of spent catalysts (EtOH/H_2O = 1/12, 420 °C). The base pressure in the analysis chamber was 4×10^{-7} Pa. Data processing was performed with CasaXPS software (v 2.3.16 PR 1.6, Zurich, Switzerland), taking into account the relative sensitivity factors (provided by CasaXPS software). The XPS spectra were charge-corrected for an O 1s peak binding energy equal to 530.0 eV. After conducting Shirley background subtraction, the quantification of the Ni 2p and Mn 2p regions was performed using the peak fitting procedure of mixed Gaussian–Lorentzian components, except for the metallic cobalt and nickel for fitting, of which the LA(1.4,5,5) and LA(1.1,2.2,10) mixed functions were used, respectively. The fitting parameters for those regions were constrained and determined from the reference spectra recorded over: Ni(0) from metallic nickel and Ni(II) from nickel(II) oxide.

Thermogravimetric (TG) analysis was conducted on a TG121 microbalance system (CAHN, Newington, NH, USA) in order to quantify the amount of carbon deposited onto the catalyst surface under SRE conditions. The studies were carried out at the temperature of 420 °C at two steam to ethanol ratios, i.e., H_2O:EtOH = 4:1 and 12:1. The total volumetric flow rate of the mixture (70 mL min^{-1}) was kept constant by adding helium. Prior to reaction, the catalyst sample (0.01 g) was reduced by passing 10% H_2/He flow at the temperature of 500 °C for 1 h.

2.3. Catalyst Evaluation in the SRE Process

Studies were performed using the application of a Microactivity Reference unit (PID Eng & Tech., Madrid, Spain), similar to that described in our earlier reports. The samples of catalysts diluted with quartz were introduced to the fixed-bed continuous-flow quartz reactor. Activation was performed in a stream of hydrogen with a flow rate of 100 mL min^{-1} at 500 °C under isothermal conditions for 1 h. After activation, the catalysts were cooled down to 420 °C, and then hydrogen was replaced by the reaction mixture composed of H_2O/EtOH/Ar = 48/4/52 vol%, H_2O/EtOH/Ar = 36/4/60 vol%, H_2O/EtOH/Ar = 24/4/72 vol%, and H_2O/EtOH/Ar = 16/4/80 vol% for H_2O/EtOH molar ratios of 12/1, 9/1, 6/1, and 4/1, respectively. The total flow rate was equal to 100 mL min^{-1} and the weight of the catalyst was equal to 100 mg. Space velocity referenced to the total flow rate of the reaction mixture divided by the catalyst weight was equal to 60,000 mL h^{-1} g^{-1} (space velocity referenced to the flow rate of EtOH was regarded as a key component of the reaction; divided by the catalyst weight, it was equal to 2400 mL $_{EtOH}$ h^{-1} g^{-1}). The reaction substrates and products were analyzed with two online gas chromatographs. One of them, Bruker 450-GC was equipped with two columns, the first filled with a porous polymer Porapak Q (for all organics, CO_2 and H_2O vapor) and the other a capillary column CP-Molsieve 5Å (for CH_4 and CO analysis). Helium was used as a carrier gas and a TCD detector was employed. The hydrogen concentration was analyzed by the second gas chromatograph, Bruker 430-GC, using a Molsieve 5Å, argon as a carrier gas, and a TCD detector. Both Bruker 450-GC and Bruker 430-GC were equipped with TCD detectors.

The conversion of ethanol (X_{EtOH}) and conversions of ethanol into individual carbon-containing products (X_{CP}) were calculated on the basis of its concentrations at the reactor inlet and outlet:

$$X_{EtOH} = \frac{C_{EtOH}^{in} - C_{EtOH}^{out}}{C_{EtOH}^{in}} \times 100\% \qquad (1)$$

$$X_{CP} = \frac{n_i C_i^{out}}{\sum n_i C_i^{out}} \times 100\% \qquad (2)$$

where C_{EtOH}^{in}-is the molar concentration of ethanol in the reaction mixture (mol%), C_{EtOH}^{out}-is the molar concentration of ethanol in the post-reaction mixture (mol%), C_i^{out}-is the molar concentration of carbon-containing product in the post-reaction mixture (mol%), and n_i–is the number of carbon atoms in the carbon-containing molecule of the reaction product.

The selectivity of hydrogen formation was determined as:

$$H_2 \text{selectivity} = \frac{C_{H_2}^{out}}{C_{H_2}^{out} + 2 \times C_{CH_4}^{out} + 2 \times C_{CH_3CHO}^{out}} \times 100\% \qquad (3)$$

where C^{out}-is the molar concentration of the hydrogen-containing products in the post-reaction mixture (mol%).

3. Results and Discussion

3.1. The Influence of a Potassium Promoter on the Catalysts' Physicochemical Properties

The chemical compositions and textural properties of MnO_x support and both nickel-based catalysts are summarized in Table 1. The results of XRF studies reveal that the assumed content of nickel is obtained with satisfactory accuracy. However, the actual weight percentage of potassium loadings in the KNi/MnO_x catalyst is less than the nominal value. It is well-known that in order to achieve high metal dispersion and high thermal stability, the support material should have a high specific surface area [35,36]. However, the BET surface area of the obtained MnO_x support was very low, which led to obtaining the catalysts with low surface areas and rather low nickel dispersion. TEM studies of the Ni/MnO_x catalyst (Table 1, see Supporting Information, Figure S1a) demonstrate that, although the Ni^0 particles have a narrow size distribution in the range of 6–24 nm, the average particle size is high and equal to 14.2 nm. Compared to the Ni/MnO$_x$ catalyst, Ni^0 particles of the KNi/MnO$_x$ sample (Table 1, see Supporting Information, Figure S2b) exhibited a wider size distribution in the range of 6–38 nm with a larger average particle size of 20.4 nm. Based on the Scherrer equation analysis of the (111) reflections, the average size of the obtained Ni^0 particles of the Ni/MnO$_x$ and KNi/MnO$_x$ catalysts was estimated to be 19.0 and 23.8 nm, respectively (Table 1). Although slightly different values of Ni^0 particle size were found for TEM and XRD methods, a trend of increasing Ni^0 particle size in the presence of a potassium promoter is the same. Probably, the aggregation of the primarily formed nickel particles of the Ni/MnO$_x$ sample during its second calcination after impregnation with potassium salt led to the formations of nickel particles with a larger size.

Table 1. Physical and chemical properties of the MnO$_x$ support and nickel-based catalysts.

Sample	Metal Content (wt.%)		S$_{BET}$ (m^2 g^{-1})	Ni^0 Particle Size (nm) *	
	Ni	K		By XRD	By TEM
Ni/MnO	9.7	-	16.7	19	14
KNi/MnO	9.1	1.6	10.1	23	20
MnO$_x$	-	-	12.0	-	-

* Ni^0 denoted Ni particle size for catalyst reduced at 500 °C.

HRTEM studies were carried out to obtain an overview of the morphology and latticed spacings of nickel-based catalysts after their reduction at 500 °C. A representative selection of HRTEM images of the Ni/MnO$_x$ and KNi/MnO$_x$ catalysts is shown in Figure 1a–b. The corresponding fast Fourier transform (FFT) pattern (Figure 1a$_1$,b$_1$) adequately verifies the formation of metallic particles of nickel in its elemental state Ni^0 and low valent manganese oxide MnO after thermal treatment of both catalysts under a reducing atmosphere. The FFT patterns indicate the characteristic (111) intensity ring and plane, with a d-spacing of 0.203 nm for face-centered cubic metallic nickel. the d-spacing of 0.222 nm correspond to the (200) lattice fringes of cubic MnO. These findings are consistent with the results obtained by X-ray diffraction studies (Figure 2c,d, see Supporting Information, Figure S2b). Recorded XRD patterns of both catalysts as-prepared and reduced with hydrogen at 500 °C confirm the structural transformation that accompanies their reduction. In the case of as-prepared samples (Figure 2a,b), the diffraction peaks at 2θ exhibited a mixed-phase structure of NiO, Mn$_2$O$_3$, and Mn$_3$O$_4$; the XRD results of the reduced samples (Figure 2c,d) indicate that the series of Bragg reflections corresponding to two crystalline phases, Ni^0 and MnO, is maintained (see Supporting Information, Table S1) [37–42]. This means that, during the treatment of both nickel-based catalysts under a hydrogen atmosphere at 500 °C, Ni^0 and MnO phases are formed by the reduction of NiO and Mn$_2$O$_3$/Mn$_3$O$_4$ phases, respectively. The comparison of XRD results for the as-prepared MnO$_x$ support (see Supporting Information, Figure S2a) and both nickel-based catalysts (Figure 2a,b) indicates a change in the structure of the MnO$_x$ phase during the step of catalyst synthesis concerning nickel introduction. The as-prepared MnO$_x$ support mainly exhibits a phase of hausmannite (Mn$_3$O$_4$) with spinel structure and tetragonal symmetry, with a minor contribution of cubic Mn$_2$O$_3$ phase, while XRD patterns of both as-prepared catalysts include a dominant Mn$_2$O$_3$ phase and a minor phase of hausmannite (Mn$_3$O$_4$) [43]. This indicates that, during the calcination of the nickel phase precursor at the temperature of 500 °C, the partial transformation of Mn$_3$O$_4$ to Mn$_2$O$_3$ occurs, which is closely related to the oxygen transfer on the surface: O_2 (gas) $\leftrightarrow$ O_2 (ads) $\leftrightarrow$ O_2^- (ads) $\leftrightarrow$ $2O^-$ (ads) $\leftrightarrow$ $2O^{2-}$ (ads) $\leftrightarrow$ $2O^{2-}$ (lattice) [44]. Because there is not much difference between the XRD patterns of the Ni/MnO$_x$ and KNi/MnO$_x$ catalysts, it can be suggested that no phase is changed after the potassium addition. Any phases containing potassium are not identified, indicating that the potassium compound remained amorphous [30] or was highly dispersed on the catalysts' surface [21]. STEM-EDS analysis (Figure 1b$_3$) confirms that potassium is well-dispersed and distributed over both MnO and Ni^0 particles of the reduced KNi/MnO$_x$ sample. On the other hand, the same analysis indicates an inhomogeneity of Ni^0 distribution on the MnO support for both the reduced Ni/MnO$_x$ (Figure 1a$_3$) and KNi/MnO$_x$ (Figure 1b$_3$) catalysts. From visual inspections of the maps, it is clear that in the case of both nickel-based catalysts, there are Ni-rich and Ni-poor regions but it is difficult to establish which nickel-based catalyst exhibited more uniformly distributed Ni^0 particles.

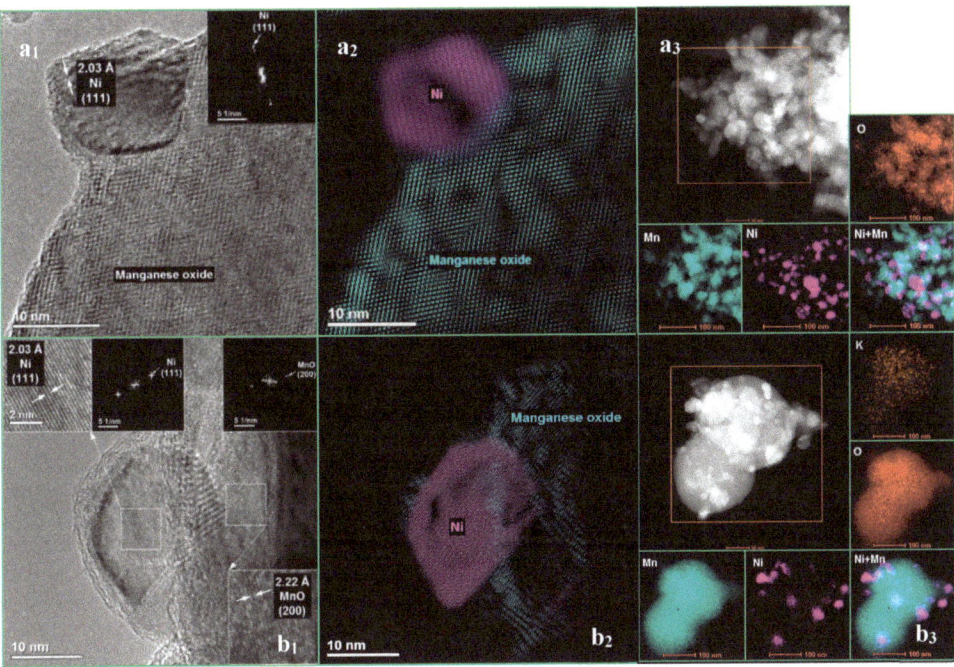

Figure 1. Microscopic analysis of the (**a**) Ni/MnO$_x$ and (**b**) KNi/MnO$_x$ catalysts after reduction at 500 °C. HRTEM images (**a$_1$,a$_2$,b$_1$,b$_2$**) and STEM-EDS analysis (**a$_3$,b$_3$**).

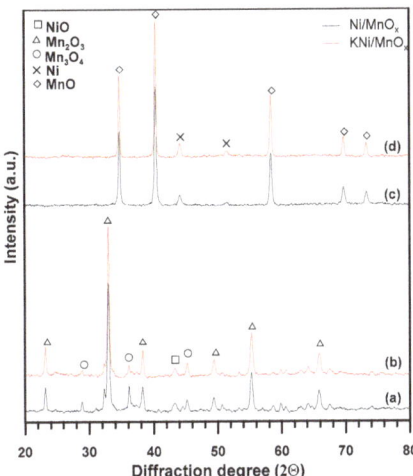

Figure 2. XRD patterns of (**a,b**) as prepared and (**c,d**) reduced with hydrogen at 500 °C for Ni/MnO$_x$ and KNi/MnO$_x$ catalysts.

A possible role of the alkali ions as promoters is that they could change the reducibility of the catalyst. Therefore, the relative reducibility of the nickel-based catalysts was investigated using an H_2-TPR measurement. Because it is known that NiO and MnO_x oxides are reduced approximately in the same temperature range [43–45] (see Supporting Information, Figure S3), the H_2-TPR profile of NiO, KNiO, MnO_x, and $KMnO_x$ samples used as the reference samples were also obtained. As seen in Figure 3, the reduction of the MnO_x support can be divided into two stages, namely a low-temperature reduction (LTR) ranging ca. 220–330 °C and a high-temperature reduction (HTR) ranging ca. 330–520 °C. The former reduction peaks can be attributed to the reduction of Mn_2O_3 to Mn_3O_4, and the later reduction peak can be ascribed to the transformation of Mn_3O_4 to MnO, which is in line with the X-ray diffraction result (see Supporting Information, Figure S2) [17,44,46,47]. Compared to MnO_x, the reduction peaks of Ni/MnO_x catalysts obviously shifted to lower temperatures, suggesting the excellent redox ability of the sample. The first main reduction peak shifted from 417 °C to 384 °C and the second main reduction peak shifted from 309 °C to 288 °C. The low-temperature peak can be ascribed to the reduction of Mn_2O_3 to Mn_3O_4, and the high-temperature peak can be ascribed to the overlap of two reduction processes: the reduction of Mn_3O_4 to MnO and the reduction of NiO to Ni^0. The promotion of MnO_x reduction to MnO by nickel can be interpreted in terms of the activation and spillover of hydrogen from the initially reduced nickel (Ni^0) to MnO_x [48,49]. Because the K-containing materials generally show a wide peak and move to a lower temperature [50], the H_2-TPR profile of the KNi/MnO_x catalyst shows only one reduction consumption peak between 160–420 °C. The main highly asymmetric peak with the maximum at 350 °C could be attributed to two overlapping processes: rapid reduction of Mn_2O_3 to MnO with Mn_3O_4 as intermediates and the reduction of NiO to Ni^0. A small shoulder peak in the low-temperature region represents the existence of unstable species with different Mn–O bond strengths, which are unstable in the lattice of oxides and could be regarded as surface-active species [50]. The peak of reduction gradually shifted to a lower value after the potassium promoter addition, indicating a weakened interaction between nickel species and the MnO_x support compared to the Ni/MnO_x sample [51]. Moreover, according to Grzybek et al. [30], the presence of potassium facilitates the storage of nitrates on the catalyst's surface. Therefore, the appearance of the intense sharp peaks in the H_2-TPR profile of KNi/MnO_x catalyst can result from the presence of a significant amount of nitrate residues accumulated on the catalyst's surface. Probably, the nickel presence increases the nitrate reducibility because hydrogen is dissociated on the reduced nickel (Ni^0) and then it spills to the nitrate ions and reduces them [52]. The reduction of nitrates explains the high hydrogen consumption obtained for the KNi/MnO_x sample (see Supporting Information, Table S2).

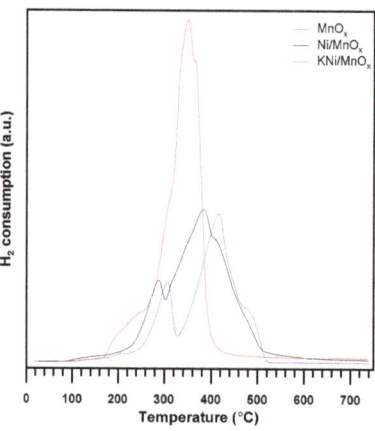

Figure 3. H_2-TPR profiles of MnO_x support and Ni/MnO_x and KNi/MnO_x catalysts.

In order to better understand the nature of the acid sites and the role of the nickel and potassium addition, in situ FT-IR experiments on MnO_x, Ni/MnO_x, and KNi/MnO_x samples were performed (Figure 4). For the MnO_x support, the IR spectra of pyridine sorption show the presence of the 1445 cm^{-1} band corresponding to Py interacting with Lewis acid sites with medium strength, which originate from the Mn(IV, III) surface exposed cations. The complexity of the PyL band at 1445 cm^{-1} points to the heterogeneous nature of Mn surface cations ruled by their various oxidation states (IV, III) and/or their coordination with O^{2-}. The deposition of nickel resulted in the appearance of an additional band at 1430 cm^{-1}, which originated from the physisorbed Py molecules. The total concentration of Lewis acid sites of the Ni/MnO_x catalyst is comparable to that of the MnO_x support (Table 2). However, their strength is higher, giving an intense peak at 1448 cm^{-1}. Morrow et al. [53] demonstrated that nitrogen-coordinated pyridine chemisorbs perpendicular to the surface on nickel-metal moieties; thus, such metal-originated Lewis acid sites can give rise to 1448 cm^{-1} IR bands [54]. Consequently, it can be concluded that the presence of metallic forms of nickel does not reduce the number of Lewis acid centers, but furthermore contributes to the increase of the strength of Lewis acid centers.

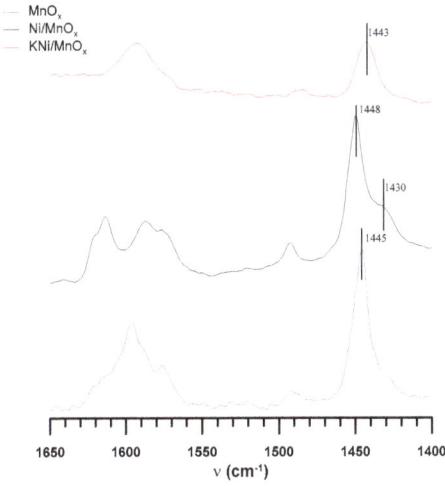

Figure 4. FT-IR spectra of pyridyne interacting with the Lewis acid sites of MnO_x support and Ni/MnO_x and KNi/MnO_x catalysts.

Table 2. The comparison of the content of Lewis acid sites for MnO_x support and Ni/MnO_x and KNi/MnO_x catalysts.

Sample	Lewis Acid Sites Concentration (mmol g^{-1})
MnO_x	54
Ni/MnO_x	50
KNi/MnO_x	13

The potassium addition to the Ni/MnO_x catalyst resulted in the downshift of this band to 1443 cm^{-1}, indicating the decrease of the electron acceptor properties of the Lewis sites, which was additionally accompanied by a significant decrease in the number of Lewis acid centers. This phenomenon is ascribed to an increase in the surface electron density induced by K-dopant. This is in line with the results of the work function studies, which show a decrease in the work function of the Ni/MnO_x catalysts upon potassium doping from 4.9 to 4.6 eV, also in accordance with the previous literature reports [55].

3.2. The Influence of a Potassium Promoter on the Performance of the Catalysts in Ethanol Steam Reforming Process

The influence of a potassium promoter on the nickel-based catalysts in the SRE process was evaluated in terms of ethanol conversion (Figure 5) and selectivity to products (Figure 6a,b) at the temperature of 420 °C for different H_2O/EtOH molar ratios of 12/1, 9/1, 6/1, and 4/1. Additionally, the performance of the MnO_x support under SRE conditions was carried out (see Supporting Information, Figure S4) and was found to be almost inactive in the SRE process with ethanol conversion lower than 5%. As expected, due to its basicity, MnO_x favors ethanol dehydrogenation to acetaldehyde [56]. The lack of any C1 compound among the products confirms an inability of MnO_x to break the C–C bond in the absence of the nickel active phase. These results are in line with those obtained for the MnO_x support performance under SRE conditions by Gac et al. [21].

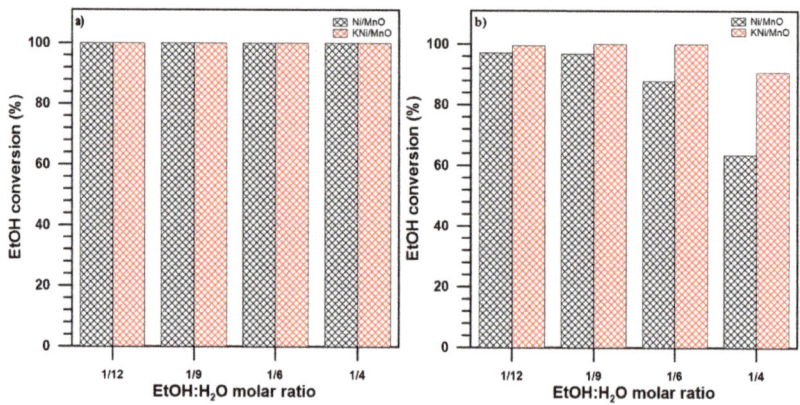

Figure 5. The ethanol conversion (**a**) at the beginning and (**b**) after 18 h of SRE process at 420 °C for H_2O/EtOH molar ratio of 12/1, 9/1, 6/1, and 4/1 over Ni/MnO_x and KNi/MnO_x catalysts.

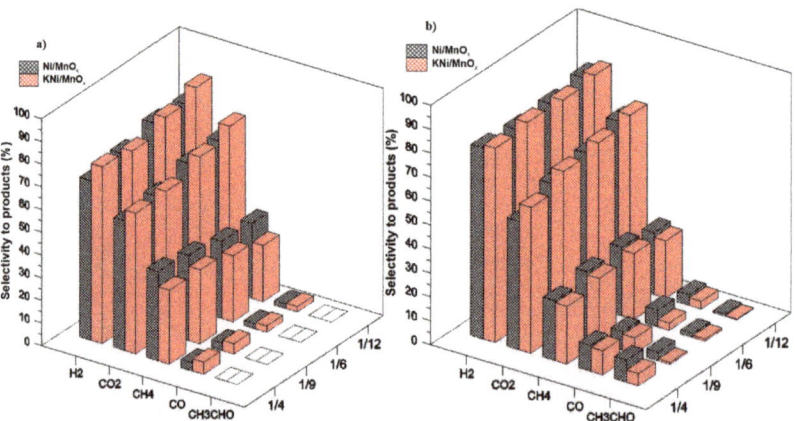

Figure 6. The selectivity to products (**a**) at the beginning and (**b**) after 18 h of SRE process at 420 °C for H_2O/EtOH molar ratio of 12/1, 9/1, 6/1, and 4/1 over Ni/MnO_x and KNi/MnO_x catalysts.

Regardless of the H_2O/EtOH molar ratio, at the beginning of the SRE process, both catalysts exhibited 100% ethanol conversion. After 18 h of processing, the ethanol conversion over the Ni/MnO_x sample decreased in the whole range of molar ratios studied because of the carbon formation on the catalyst surface (the studies on catalyst deactivation will be discussed in the next section). The higher excess of water facilitates carbon gasification [56].

Therefore, the highest ethanol conversion of 97% after 18 h of the SRE process was observed for the H_2O/EtOH molar ratio of 12/1 and it decreased with the decrease of excess water to 63% for the H_2O/EtOH molar ratio of 4/1. For the KNi/MnO_x catalyst, complete ethanol conversion was obtained for 18 h of SRE reaction in the range of H_2O/EtOH molar ratio of 6/1–12/1. The decrease of the sample activity to ca. 90% after 18 h of SRE reaction was only observed at the low H_2O/EtOH molar ratio of 4/1. The obtained results indicate that the potassium promoter inhibits catalyst deactivations, thus improving its stability under SRE conditions (see Supporting Information, Figure S5), which is consistent with the previous studies on the role of this catalyst promoter in the studied process [25,26,29,34,57,58]. On the other hand, the stability of nickel-based catalysts in the SRE reaction is not always improved by the addition of potassium, which was shown by Słowik et al. [28].

The distribution of products over both nickel-based catalysts indicates that, besides the SRE reaction (Reaction (R1)), ethanol decomposition (Reaction (R2)) and its dehydrogenation (Reaction (R3)) followed by acetaldehyde decomposition (Reaction (R4)) also occur.

$$C_2H_5OH + 3H_2O \leftrightarrow 2CO_2 + 6H_2 \quad (R1)$$

$$C_2H_5OH \leftrightarrow CH_4 + CO + H_2 \quad (R2)$$

$$C_2H_5OH \leftrightarrow CH_3CHO + H_2 \quad (R3)$$

$$CH_3CHO \leftrightarrow CH_4 + CO \quad (R4)$$

At the beginning of the SRE reaction (Figure 6a), regardless of H_2O/EtOH molar ratio, both Ni/MnO_x and KNi/MnO_x catalysts exhibit the highest selectivity to hydrogen and carbon dioxide, suggesting that the SRE is the main reaction taking place (Reaction (R1)). More of the most desirable products of SRE reaction is produced over KNi/MnO_x catalyst which could be ascribed to increase the activity of SRE (Reaction (R1)) and the water gas shift (WGS, Reaction (R5)) reactions in the presence of potassium promoted sample.

$$CO + H_2O \leftrightarrow CO_2 + H_2 \quad (R5)$$

Moreover, the rate of both SRE (Reaction (R1)) and WGS (Reaction (R5)) reactions increases with the increase in steam excess [59]. Therefore, in the case of both nickel-based catalysts, the selectivity to carbon dioxide is the lowest for the H_2O/EtOH molar ratio of 4/1 (57% for Ni/MnO_x, 62% for KNi/MnO_x) but increases with its increase to 12/1 (65% for Ni/MnO_x and 73% for KNi/MnO_x). At the beginning of the SRE reaction, carbon monoxide and methane are only byproducts that are formed under SRE conditions. In the whole range of H_2O/EtOH molar ratios studied, the selectivity to carbon monoxide does not exceed 3.5 and 5% over the Ni/MnO_x and KNi/MnO_x catalysts, respectively. However, selectivity to the second byproduct, methane, is much higher in the presence of both materials because of the high activity of nickel-based catalysts in the methanation reactions of carbon monoxide and/or carbon dioxide (Reactions (R6) and (R7)) [60].

$$CO + 3H_2 \leftrightarrow CH_4 + H_2O \quad (R6)$$

$$CO_2 + 4H_2 \leftrightarrow CH_4 + 2H_2O \quad (R7)$$

The decrease in selectivity to methane from 36 to 33% (Ni/MnO_x) and from 32 to 25% (KNi/MnO_x) with the increase in the H_2O/EtOH molar ratio from 4/1 to 12/1 confirms that excess of water facilitates the steam reforming of methane (Reaction (R8)) [60].

$$CH_4 + H_2O \leftrightarrow CO + 3H_2 \quad (R8)$$

After 18 h of the SRE reaction (Figure 6b), the distribution of products in the presence of both nickel-based catalysts is comparable to that observed at the beginning of the process for a whole range of H_2O/EtOH molar ratios studied. The sole appearance of acetaldehyde among byproducts after 18 h of the SRE reaction indicates a decrease in the ability of both catalysts in the C–C bond cleavage with the increase in time-on-stream due to carbon formation. Regardless of the $H_2O/EtOH$ molar ratio, more nickel active sites to break this bond remained available in the case of the potassium-promoted catalyst, resulting in its lower selectivity to the C2 product, i.e., acetaldehyde, and higher selectivity to the C1 products, i.e., carbon monoxide, carbon dioxide, and methane, in comparison with the Ni/MnO_x sample. Because steam excess limits the carbon formation, the amount of produced acetaldehyde increases with a decrease in $H_2O/EtOH$ molar ratio.

3.3. The Influence of a Potassium Promoter on Prevention of the Nickel-Based Catalyst Deactivation under SRE Conditions

The SEM (Figures 7a$_1$ and 8a$_1$) and TEM images (Figures 7a$_2$,a$_3$ and 8a$_2$,a$_3$) of the Ni/MnO_x and KNi/MnO_x catalysts show that carbon filamentous deposits are formed on the surface of both samples after 18 h of the SRE reaction at 420 °C, regardless of the $H_2O/EtOH$ molar ratio. Carbon filaments, leading to catalyst deactivation, are typically formed at the surface of nickel particles under SRE conditions via the Boudouard reaction (Reaction (R9)) and/or methane decomposition [51].

$$2CO \leftrightarrow C + CO_2 \tag{R9}$$

$$CH_4 \leftrightarrow C + 2H_2 \tag{R10}$$

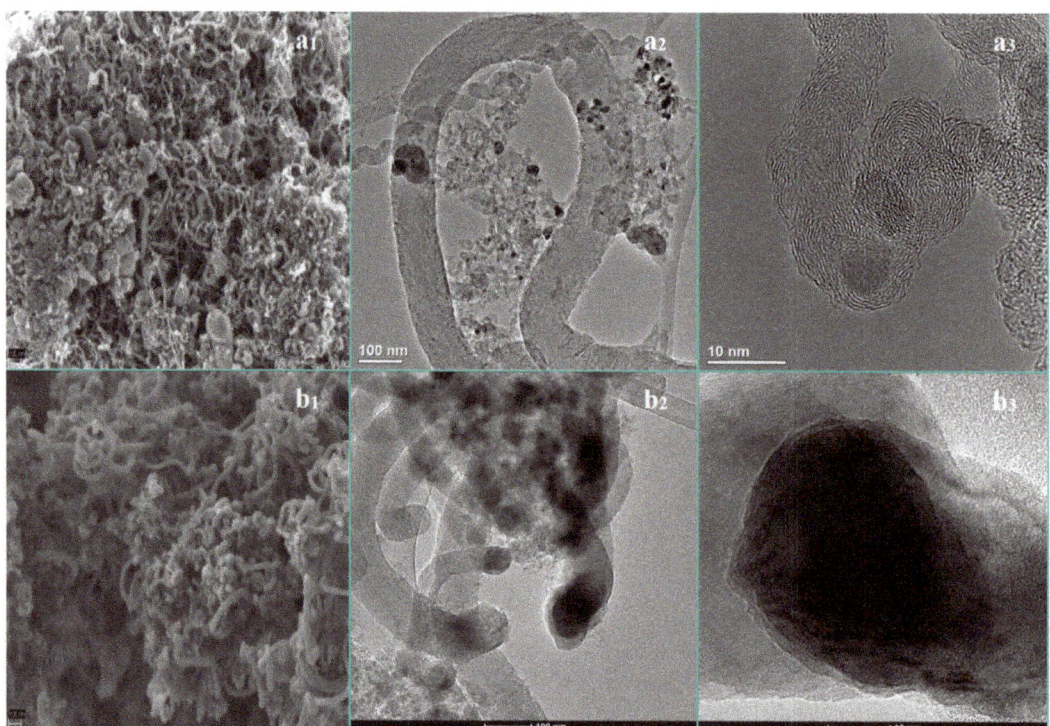

Figure 7. Microscopic analysis of Ni/MnO_x catalyst after 18 h of SRE reaction at 420 °C for (**a**) $H_2O/EtOH$ molar ratio of 12/1 and (**b**) $H_2O/EtOH$ molar ratio of 4/1. SEM images (**a$_1$,b$_1$**) and TEM images (**a$_2$,a$_3$,b$_2$,b$_3$**).

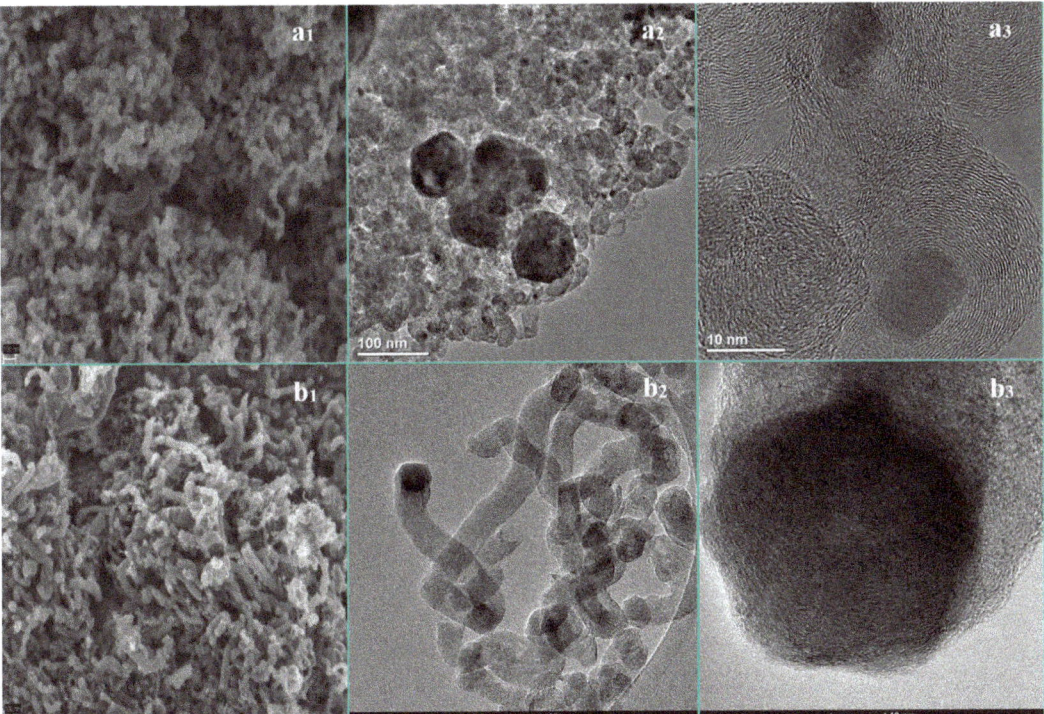

Figure 8. Microscopic analysis of KNi/MnO$_x$ catalyst after 18 h of SRE reaction at 420 °C for (**a**) H$_2$O/EtOH molar ratio of 12/1 and (**b**) H$_2$O/EtOH molar ratio of 4/1. SEM images (**a$_1$**,**b$_1$**) and TEM images (**a$_2$**,**a$_3$**,**b$_2$**,**b$_3$**).

Therefore, comparing SEM images of spent catalysts and support (see Supporting Information, Figure S6) reveals that carbon deposition occurs efficiently on nickel surface due to its ability to break C–C and C–H bonds, while relatively little (or lack) carbon deposition occurs on MnO$_x$ without nickel. Since the size distribution of the nickel particles was not in a narrow range for both Ni/MnO$_x$ and KNi/MnO$_x$ catalysts (see Supporting Information, Figure S1), the carbon filaments exhibited a correspondingly non-uniform diameter distribution. Moreover, carbon filaments with not only significantly varying diameters but also lengths formed on the surface of both nickel-based catalysts under SRE conditions. The presence of potassium seems not to influence both diameters and lengths of the carbon filaments. However, the degree of carbon graphitization depends on the potassium promotion, as can be seen in the TEM images (Figures 7a$_3$ and 8a$_3$). The presence of potassium influenced the increase in the degree of graphitization and a highly ordered, i.e., graphitic carbon formed on the surface of the KNi/MnO$_x$ catalyst, whereas mixed turbostratic carbon and graphitic carbon phases formed on the surface of the Ni/MnO$_x$ sample. These results are consistent with those obtained by Słowik et al. [28], who observed an increase in the degree of graphitization of carbon formed under SRE conditions over KNi/CeO$_2$ compared to the Ni/CeO$_2$ catalyst.

By comparing spatial features in Ni and Mn maps of both catalysts after the SRE reaction (Figures 9a$_5$ and 10a$_5$), it is apparent that there are regions with the presence of Ni but without Mn copresence. This means that the Ni particles did not remain attached to the support surface but a carbon deposit separated the Ni phase from the support. Upon closer examination (Figures 7a$_3$,b$_3$, 8a$_3$,b$_3$, 9a$_1$–a$_5$ and 10a$_1$–a$_5$), it was evident that the Ni particles were carried away from the support surface by the carbon growth process during the reaction and they were located at the tip of the filaments, which is a direct consequence of the weak metal-support interaction(s) associated with both Ni/MnO$_x$ and KNi/MnO$_x$ catalysts. In contrast to these results, where a potassium promotion does not enhance the interaction between Ni and MnO$_x$ phases, the strong cobalt-support interaction resulting from the potassium presence resulted in the stabilization of the active phase, preventing its removal from the Al$_2$O$_3$ surface because of carbon filament growth under SRE conditions [30].

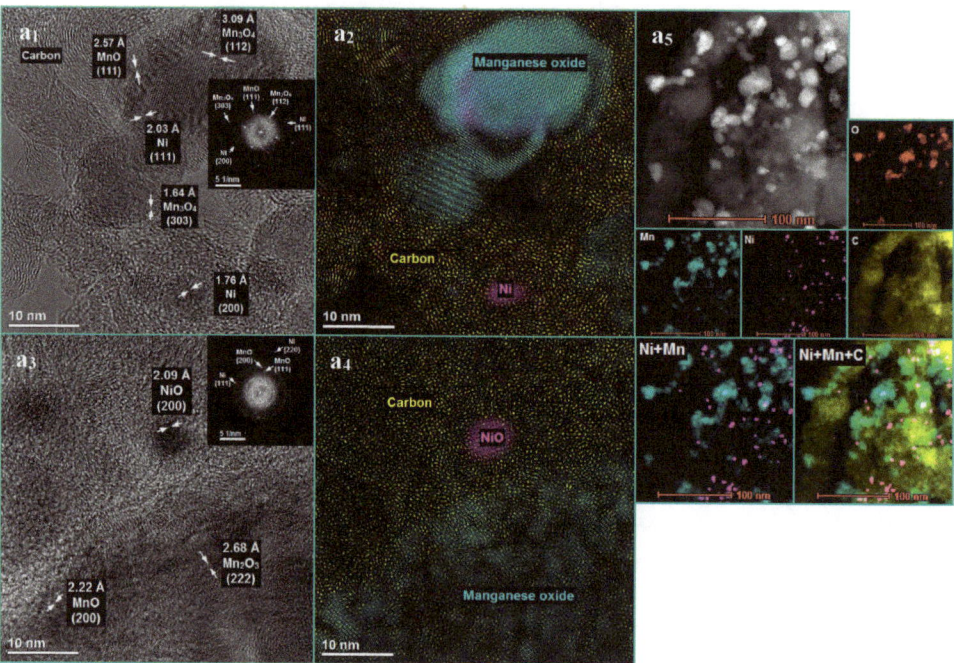

Figure 9. Microscopic analysis of the Ni/MnO$_x$ catalyst after 18 h of SRE reaction at 420 °C (H$_2$O/EtOH molar ratio = 12/1). HRTEM images (a$_1$, a$_2$, a$_3$, a$_4$) and STEM-EDS analysis (a$_5$).

The TEM analysis of both nickel-based samples after the SRE process at the H$_2$O:EtOH molar ratio of 12/1 indicates modifications in the distribution of nickel particle size compared to results obtained for the catalysts after their reduction (see Supporting Information, Figures S1 and S7). A comparison of histograms of the reduced samples with those obtained for catalysts after the SRE reaction shows that the larger crystallites mostly disappeared. The average nickel particle size underwent ~50% reduction (Table 3) after the SRE process at the H$_2$O:EtOH molar ratio of 12/1. This suggests that the fragmentation of the initial metal surface occurs prior to the growth of carbon filaments. On the other hand, the nickel particle size distribution obtained for nickel-based catalysts after the SRE reaction at the H$_2$O:EtOH molar ratio of 4/1 was broader and shifted to larger particles (5–65 nm) (see Supporting Information, Figure S8) in comparison with results obtained for them after reduction (see Supporting Information, Figure S1). The increase of particle size after the SRE reaction at the H$_2$O:EtOH molar ratio of 4/1 results from sintering. However, the number of large crystallites is not high and the average particle size (Table 2) is comparable with those obtained for reduced catalysts. The mainly small crystallites were observed after the SRE reaction at the H$_2$O:EtOH molar ratio of 4/1, suggesting that nickel crystallites underwent a continued fragmentation into smaller ones.

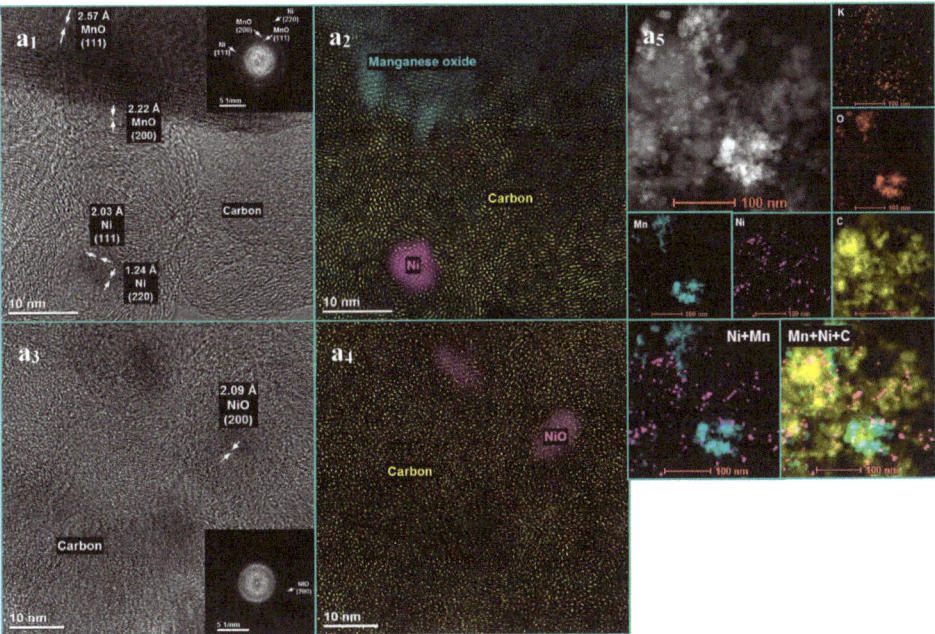

Figure 10. Microscopic analysis of the KNi/MnO$_x$ catalyst after 18 h of SRE reaction at 420 °C (H$_2$O/EtOH molar ratio = 12/1). HRTEM images (a$_1$,a$_2$,a$_3$,a$_4$) and STEM-EDS analysis (a$_5$).

Table 3. Ni particle size of the Ni/MnO$_x$ and KNi/MnO$_x$ catalysts after SRE reaction.

Sample	Ni Particle Size (nm)	
	H$_2$O:EtOH = 12/1	H$_2$O:EtOH = 4/1
Ni/MnO	8	18
KNi/MnO	10	20

The FFT method (Figures 9a$_1$,a$_3$ and 10a$_1$,a$_3$) reveals the presence of both Ni^0 and NiO phases in both spent catalysts. Moreover, based on the phase identification by the FFT method, the MnO$_x$ support of the spent Ni/MnO$_x$ sample was identified to include three forms: MnO, Mn$_2$O$_3$, and Mn$_3$O$_4$, where only facets originating from MnO were detected for the support of the spent KNi/MnO$_x$ catalyst. Moreover, the surface composition was investigated by XPS spectroscopy. The surface of both spent nickel-based catalysts was completely covered by carbon (96 at.%) and oxygen to carbon-bounded species (3.9 at.%); therefore, the intensity of the Mn 2p and Ni 2p regions was very low, almost at the level of noise (Figure 11). Due to this fact, the fitting procedure and calculation of each component's contribution to the overall spectrum were affected by a significant error and will not be discussed in this work. However, it is worth seeing that, on the surface of both catalysts, two oxidation states of manganese and nickel are recorded. The Mn 2p region is slightly shifted towards lower binding energies, suggesting that most manganese may exist as MnO [61]. It should be highlighted that the Mn 2p region overlaps with Ni LMM Auger peaks [62]. In order to analyze the contribution of Ni LMM Auger in overall Mn 2p, we recorded Ni LMM template from Ni0 and NiO. Using these lines and the calculated area ratio between Ni 2p3/2/Ni LMM for Ni0 and Ni 2p 3/2/Ni LMM for NiO, we estimated the contribution of Ni LMM in Mn 2p3/2. Since fitted signals have very small intensity, they were omitted in further analysis of this region. In the case of nickel, the shape of the recorded spectra undoubtedly indicates the presence of NiO (pronounced shoulder at

855.2 eV), whereas the peak at 852.5 eV is related to the metallic species [61,63]. The results obtained from both FFT and XPS methods indicate that both the pre-reduced Ni^0 active phase and MnO support of the Ni/MnO$_x$ and KNi/MnO$_x$ samples are oxidized under SRE conditions. However, neither of these methods allow us to determine whether potassium influences the oxidation state of the surface during this process.

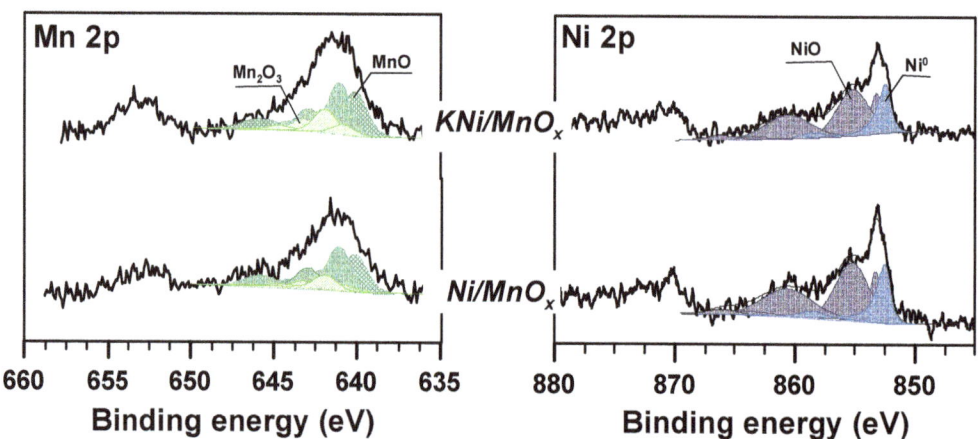

Figure 11. High-resolution XPS spectra of Mn 2p and Ni 2p regions collected from the surface of Ni/MnO$_x$ and KNi/MnO$_x$ catalysts after 18 h of SRE reaction at 420 °C (H$_2$O/EtOH molar ratio = 12/1).

To study the difference in the number of accumulated species on the Ni/MnO$_x$ and KNi/MnO$_x$ catalysts and the rate of their formation, the thermogravimetric experiments under SRE conditions were conducted for the H$_2$O/EtOH molar ratio of 12/1 and 4/1 at 420 °C (Figure 12). The results from the thermogravimetric experiments show that the amount of carbon deposits for both nickel-based catalysts is minimized by performing the SRE reaction with the higher excess of water. Moreover, regardless of the H$_2$O/EtOH molar ratio, the rate of carbon formation is significantly lower over KNi/MnO$_x$ compared to Ni/MnO$_x$, which indicates that the addition of an alkali promoter inhibits carbon deposition and/or promotes its gasification. According to the literature [64–66], the presence of potassium on the Ni surface can decrease the number of active sites available for methane decomposition (reaction 10), which allows for suppression of the amount of carbon deposits. The presence of potassium dopant promotes the adsorption of water in a dissociative manner [65,66] and raises the oxygen population on the surface, which, in turn, increases the rate of the steam gasification of carbon formed during steam reforming reactions. Furthermore, the lower accumulation of the carbon deposits on the surface of the KNi/MnO$_x$ catalyst guarantees more Ni active sites available to the reactants and the higher stability of this catalyst compared to the Ni/MnO$_x$ sample.

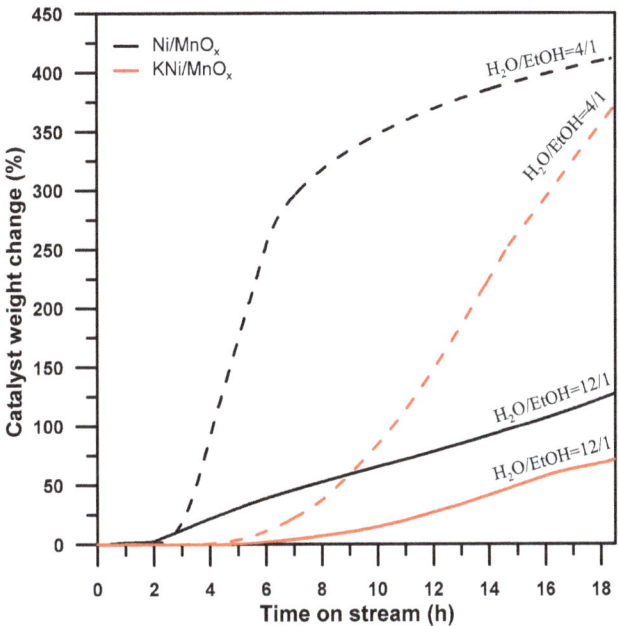

Figure 12. Changes in Ni/MnO$_x$ and KNi/MnO$_x$ catalysts' weight under SRE conditions at H$_2$O/EtOH molar ratio = 12/1 (straight line) and H$_2$O/EtOH molar ratio = 4/1 (dashed line) at 420 °C.

4. Conclusions

By modifying Ni/MnO$_x$ with a potassium promoter, the catalyst stability can be improved under the SRE conditions for different H$_2$O/EtOH molar ratios in the range from 4/1 to 12/1. Moreover, since the KNi/MnO$_x$ catalyst is more resistant to deactivation, more nickel active sites to break the C–C bond remain available on its surface, resulting in its lower selectivity to acetaldehyde and higher selectivity to C1 products in comparison to the Ni/MnO$_x$ sample after 18 h of the SRE reaction regardless of the H$_2$O/EtOH molar ratio. The observed high stability of the catalyst promoted with the potassium results from an improved resistance to carbon formation. The potassium promoter inhibits the accumulation of the carbon on the catalyst surface under SRE conditions. Because the Ni active phase dispersion over the KNi/MnO$_x$ sample is worse compared to the Ni/MnO$_x$ catalyst, the observed higher ability to inhibit the carbon formation of potassium-promoted catalysts does not result from the smaller size of Ni crystallites. It means that the restriction of the amount of carbon deposits originates from the potassium presence on the Ni surface, leading (i) to a decrease in the number of active sites available for methane decomposition and (ii) to an increase in the rate of the steam gasification of carbon formed during SRE reactions. On the other hand, the potassium addition does not influence the type of carbon deposits and filamentous carbon is formed in the presence of both Ni/MnO$_x$ and KNi/MnO$_x$ catalysts. Additionally, the morphology, i.e., thickness and length, of carbon filaments formed on the surface of both nickel-based catalysts are similar. Only the degree of graphitization of the carbon deposit on the surface of the Ni/MnO$_x$ and KNi/MnO$_x$ catalysts is slightly different. The Ni/MnO$_x$ catalyst exhibits less graphitization degree than carbon deposited on the surface of KNi/MnO$_x$ material.

Supplementary Materials: The following supporting information can be downloaded at: https://www.mdpi.com/article/10.3390/catal12060600/s1, Figure S1: Particle size distribution histograms for (a) Ni/MnO$_x$ and (b) KNi/MnO$_x$ catalysts reduced at 500 °C. Size distribution of particles determined from measurement of at least 200 particles from representative HRTEM images; Figure S2: XRD patterns of (a) calcined and (b) reduced at 500 °C of MnO$_x$ support; Figure S3: H$_2$-TPR profiles of NiO and K/NiO samples; Figure S4: Ethanol conversion and selectivity to products over MnO$_x$ support in SRE process at 420 °C. (H$_2$O/EtOH = 12/1); Figure S5: Stability tests of Ni/MnO$_x$ and KNi/MnO$_x$ catalysts at temperature of 420 °C under SRE conditions for H$_2$O/EtOH molar ratio of (a) 12/1 (b) 9/1, (c) 6/1 and (d) 4/1; Figure S6: SEM images of MnO$_x$ support (a) in the fresh state and (b) after 18 hours of SRE reaction at 420 °C (H$_2$O/EtOH molar ratio = 12/1); Figure S7: Particle size distribution histograms for (a) Ni/MnO$_x$ and (b) KNi/MnO$_x$ catalysts after SRE reaction at temperature of 420 °C under SRE conditions for H$_2$O/EtOH molar ratio of 12/1. Size distribution of particles determined from measurement of at least 200 particles from representative HRTEM images; Figure S8: Particle size distribution histograms for (a) Ni/MnO$_x$ and (b) KNi/MnO$_x$ catalysts after SRE reaction at temperature of 420 °C under SRE conditions for H$_2$O/EtOH molar ratio of 4/1. Size distribution of particles determined from measurement of at least 200 particles from representative HRTEM images; Table S1: XRD details with 2θ, and hkl values of the obtained crystalline phases for as-prepared and reduced with hydrogen at 500 °C of Ni/MnO$_x$ and KNi/MnO$_x$ catalysts [37–42]; Table S2: Hydrogen consumption from H$_2$-TPR results.

Author Contributions: Conceptualization, M.G.; Formal analysis, M.G.; Funding acquisition, M.G.; Investigation, M.G., M.R., G.S., S.T.-S., G.G. and K.G.-M.; Methodology, M.G., M.R., G.S., S.T.-S., G.G., K.G.-M. and A.K.; Writing—original draft, M.G. All authors have read and agreed to the published version of the manuscript.

Funding: This work was supported by the National Science Centre, Poland within the SONATA program under grant no. 2015/19/D/ST5/01931.

Acknowledgments: The study was carried out with equipment purchased thanks to the financial support of the European Regional Development Fund in the framework of the Polish Innovation Economy Operational Program (contract no. POIG.02.01.00-06-024/09 Centre of Functional Nanomaterials; www.cnf.umcs.lublin.pl accessed on 29 April 2022). The SEM and XPS studies of catalysts used in SRE process were carried out at the Biological and Chemical Research Centre, University of Warsaw, established within the project co-financed by the European Union from the European Regional Development Fund under the Operational Programme Innovative Economy, 2007–2013 and by the National Center for Research and Development within the Panda 2 program (contract no.501-D312-56-0000002).

Conflicts of Interest: The authors declare no conflict of interest.

References

1. Jo, W.J.; Im, Y.; Do, J.Y.; Park, N.-K.; Lee, T.J.; Lee, S.T.; Cha, M.S.; Jeon, M.-K.; Kang, M. Synergies between Ni, Co, and Mn ions in trimetallic Ni$_{1-x}$Co$_x$MnO$_4$ catalysts for effective hydrogen production from propane steam reforming. *Renew. Energy* **2017**, *113*, 248–256. [CrossRef]
2. Da Costa-Serra, J.F.; Chica, A. Catalysts based on Co-Birnessite and Co-Todorokite for the efficient production of hydrogen by ethanol steam reforming. *Int. J. Hydrogen Energy* **2018**, *43*, 16859–16865. [CrossRef]
3. Manfro, R.L.; Ribeiro, N.F.P.; Souza, M.M.V.M. Production of hydrogen from steam reforming of grycerol using nickel catalysts supported on Al$_2$O$_3$, CeO$_2$ and ZrO$_2$. *Catal. Sustain. Energy* **2013**, *1*, 60–70.
4. Marcos, F.C.F.; Lucrédio, A.F.; Assaf, E.M. Effects of adding basic oxides of La and/or Ce to SiO$_2$-supported Co catalysts for ethanol steam reforming. *RSC Adv.* **2014**, *4*, 43839–43849. [CrossRef]
5. Li, Y.; Zhang, Z.; Jia, P.; Dong, D.; Wang, Y.; Song, H.; Xiang, J.; Liu, Q.; Hu, X. Ethanol steam reforming over cobalt catalysts: Effect of range of additives on the catalytic behaviors. *J. Energy Inst.* **2020**, *93*, 165–184. [CrossRef]
6. Riani, P.; Garbarino, G.; Canepa, F.; Busca, G. Cobalt nanoparticles mechanically deposited on α-Al$_2$O$_3$: A competitive catalyst for the production of hydrogen through ethanol steam reforming. *J. Chem. Technol. Biotechnol.* **2019**, *94*, 538–546. [CrossRef]
7. Yu, S.-W.; Huang, H.-H.; Tang, C.-W.; Wang, C.-B. The effect of accessible oxygen over Co$_3$O$_4$-CeO$_2$ catalysts on the steam reforming of ethanol. *Int. J. Hydrogen Energy* **2014**, *39*, 20700–20711. [CrossRef]
8. Ibrahim, S.h.M.; El-Shobaky, G.A.; Mohamed, G.M.; Hassan, N.A. Effects of ZnO and MoO$_3$ Doping on surface and catalytic properties of manganese oxide supported on alumina system. *Open Catal. J.* **2011**, *4*, 27–35. [CrossRef]
9. Comas, J.; Mariño, F.; Laborde, M.; Amadeo, N. Bio-ethanol steam reforming on Ni/Al$_2$O$_3$ catalyst. *Chem. Eng. J.* **2004**, *98*, 61–68. [CrossRef]

10. Bshish, A.; Yaakob, A.; Ebshish, A.; Alhasan, F.H. Hydrogen production via ethanol steam reforming over Ni/Al$_2$O$_3$ catalysts: Effect of Ni loading. *J. Energy Resour. Technol.* **2014**, *136*, 012601. [CrossRef]
11. Alberton, A.L.; Souza, M.M.V.M.; Schmal, M. Carbon formation and its influence on ethanol steam reforming over Ni/Al$_2$O$_3$ catalysts. *Catal. Today* **2007**, *123*, 257–264. [CrossRef]
12. Garcia, S.R.; Assaf, J.M. Effect of the preparation method on Co/Al$_2$O$_3$ catalyst applied to ethanol steam reforming reaction production of hydrogen. *Mod. Res. Catal.* **2012**, *1*, 52–57. [CrossRef]
13. Lucredio, A.F.; Bellido, J.D.A.; Zawadzki, A.; Assaf, E.M. Co catalysts supported on SiO$_2$ and γ-Al$_2$O$_3$ applied to ethanol steam reforming: Effect of the solvent used in the catalyst preparation method. *Fuel* **2011**, *90*, 1424–1430. [CrossRef]
14. Kaddouri, A.; Mazzocchia, C. A study of the influence of the synthesis conditions upon the catalytic properties of Co/SiO$_2$ or Co/Al$_2$O$_3$ catalysts used for ethanol steam reforming. *Catal. Commun.* **2014**, *5*, 339–345. [CrossRef]
15. Carvalho, F.L.S.; Asencios, Y.J.O.; Rego, A.M.B.; Assaf, E.M. Hydrogen production by steam reforming of ethanol over Co$_3$O$_4$/La$_2$O$_3$/CeO$_2$ catalysts synthesized by one-step polymerization method. *Appl. Catal. A Gen.* **2014**, *483*, 52–62. [CrossRef]
16. Arena, F.; Torre, T.; Raimondo, C.; Parmaliana, A. Structure and redox properties of bulk and supported manganese oxide catalysts. *Phys. Chem. Chem. Phys.* **2001**, *3*, 1911–1917. [CrossRef]
17. Li, J.; Li, L.; Cheng, W.; Wu, F.; Lu, X.; Li, Z. Controlled synthesis of diverse manganese based catalysts for complete oxidation of toluene and carbon monoxide. *Chem. Eng. J.* **2014**, *22*, 59–67. [CrossRef]
18. Lee, G.; Kim, D.; Kwak, B.S.; Kang, M. Hydrogen rich production by ethanol steam reforming reaction over Mn/Co$_{10}$Si$_{90}$MCM-48 catalysts. *Catal. Today* **2014**, *232*, 139–150. [CrossRef]
19. Kwak, B.S.; Lee, G.; Park, S.-M.; Kang, M. Effect of MnO$_x$ in the catalytic stabilization of Co$_2$MnO$_4$ spinel during the ethanol steam reforming reaction. *Appl. Catal. A Gen.* **2015**, *503*, 165–175. [CrossRef]
20. Fuertes, A.; Da Costa-Serra, J.F.; Chica, A. New catalysts based on Ni-Birnessite and Ni-Todorokite for the efficient production of hydrogen by bioethanol steam reforming. *Energy Procedia* **2012**, *29*, 181–191. [CrossRef]
21. Gac, W.; Greluk, M.; Słowik, G.; Turczyniak-Surdacka, S. Structural and surface changes of cobalt modified manganese oxide during activation and ethanol steam reforming reaction. *Appl. Surf. Sci.* **2018**, *440*, 1047–1062. [CrossRef]
22. Sohrabi, S.; Irankhah, A. Synthesis, characterization, and catalytic activity of Ni/CeMnO$_2$ catalysts promoted by copper, cobalt, potassium and iron for ethanol steam reforming. *Int. J. Hydrogen Energy* **2021**, *46*, 12846–12856. [CrossRef]
23. Greluk, M.; Rybak, P.; Słowik, G.; Rotko, M.; Machocki, A. Comparative study on steam and oxidative steam reforming of ethanol over 2KCo/ZrO$_2$ catalyst. *Catal. Today* **2015**, *242*, 50–59. [CrossRef]
24. Banach, B.; Machocki, A. Effect of potassium addition on a long term performance of Co-ZnO-Al$_2$O$_3$ catalysts in the low-temperature steam reforming of ethanol: Co-precipitation vs citrate method of catalysts synthesis. *Appl. Catal. A Gen.* **2015**, *505*, 173–182. [CrossRef]
25. Greluk, M.; Rotko, M.; Machocki, A. Conversion of ethanol over Co/CeO$_2$ and KCo/CeO$_2$ catalysts for hydrogen production. *Catal. Lett.* **2016**, *146*, 163–173. [CrossRef]
26. Słowik, G.; Greluk, M.; Machocki, A. Microscopic characterization of changes in the structure of KCo/CeO$_2$ catalyst used in the steam reforming of ethanol. *Mater. Chem. Phys.* **2016**, *173*, 219–237. [CrossRef]
27. Greluk, M.; Słowik, G.; Rotko, M.; Machocki, A. Steam reforming and oxidative steam reforming of ethanol over PtKCo/CeO$_2$ catalyst. *Fuel* **2016**, *183*, 518–530. [CrossRef]
28. Słowik, G.; Greluk, M.; Rotko, M.; Machocki, A. Evolution of the structure of unpromoted and potassium-promoted ceria-supported nickel catalysts in the steam reforming of ethanol. *Appl. Catal. B Environ.* **2018**, *221*, 490–509. [CrossRef]
29. Grzybek, G.; Greluk, M.; Indyka, P.; Góra-Marek, K.; Legutko, P.; Słowik, G.; Turczyniak-Surdacka, S.; Rotko, M.; Sojka, Z.; Kotarba, A. Cobalt catalyst for steam reforming of ethanol–Insights into the promotional role of potassium. *Int. J. Hydrogen Energy* **2020**, *45*, 22658–22673. [CrossRef]
30. Grzybek, G.; Góra-Marek, K.; Patulski, P.; Greluk, M.; Rotko, M.; Słowik, G.; Kotarba, A. Optimization of the potassium promotion of the Co|α-Al$_2$O$_3$ catalyst for the effective hydrogen production via ethanol steam reforming. *Appl. Catal. A Gen.* **2021**, *614*, 1188051. [CrossRef]
31. Llorca, J.; Homs, N.; Sales, J.; Fierro, J.L.G.; Ramirez de la Piscina, P. Effect of sodium addition on the performance of Co–ZnO-based catalysts for hydrogen production from bioethanol. *J. Catal.* **2004**, *222*, 470–480. [CrossRef]
32. Oktaviano, H.S.; Trisunaryanti, W. Sol-gel derived Co and Ni based catalysts: Application for steam reforming of ethanol. *Indones. J. Chem.* **2008**, *8*, 47–53. [CrossRef]
33. Vizcaíno, A.J.; Carrero, A.; Calles, J.A. Ethanol steam reforming on Mg- and Ca-modified Cu–Ni/SBA-15 catalysts. *Catal. Today* **2009**, *146*, 63–70. [CrossRef]
34. Ogo, S.; Shimizu, T.; Nakazawa, Y.; Mukawa, K.; Mukai, D.; Sekine, Y. Steam reforming of ethanol over K promoted Co catalyst. *Appl. Catal. A Gen.* **2015**, *495*, 30–38. [CrossRef]
35. Barrientos, J.; Gonzalez, N.; Boutonnet, M.; Järås, S. Deactivation of Ni/γ-Al$_2$O$_3$ catalysts in CO methanation: Effect of Zr, Mg, Ba and Ca oxide promoters. *Top. Catal.* **2017**, *60*, 1276–1284. [CrossRef]
36. Wang, A.; Tian, Z.; Liu, Q.; Qiao, Y.; Tian, Y. Facile preparation of a Ni/MgAl$_2$O$_4$ catalyst with high surface area: Enhancement in activity and stability for CO methanation. *Main Group Met. Chem.* **2018**, *41*, 73–89. [CrossRef]
37. Richardson, J.T.; Scates, R.; Twigg, M.V. X-ray diffraction study of nickel oxide reduction by hydrogen. *Appl. Catal. A Gen.* **2003**, *246*, 137–150. [CrossRef]

38. Sharrouf, M.; Awad, R.; Roumié, M.; Marhaba, S. Structural, optical and room temperature magnetic study of Mn_2O_3 nanoparticles. *Mater. Sci. Appl.* **2015**, *6*, 850–859.
39. Khalaji, A.D.; Soleymanifard, M.; Jarosova, M.; Machek, P. Facile synthesis and characterization of Mn_3O_4, Co_3O_4, and NiO. *Acta Phys. Pol.* **2002**, *137*, 1043–1045. [CrossRef]
40. Zhang, P.; Zhan, Y.; Cai, B.; Hao, C.; Wang, J.; Liu, C.; Meng, Z.; Yin, Z.; Chen, Q. Shape-controlled synthesis of Mn_3O_4 nanocrystals and their catalysis of the degradation of Methylene Blue. *Nano Res.* **2010**, *3*, 235–243. [CrossRef]
41. Martinez de la Torre, C.; Bennewitz, M.F. Manganese oxide nanoparticle synthesis by thermal decomposition of manganese(II) acetylacetonate. *J. Vis. Exp.* **2020**, *160*, e61572. [CrossRef] [PubMed]
42. Indra, A.; Menezes, P.W.; Schuster, F.; Driess, M. Significant role of Mn(III) sites in e_g^1 configuration in manganese oxide catalysts for efficient artificial water oxidation. *J. Photochem. Photobiol. B Biol.* **2015**, *152*, 156–161. [CrossRef] [PubMed]
43. Bayón, A.; de la Peña O'Shea, V.; Coronado, J.M.; Serrano, D.P. Role of the physicochemical properties of hausmannite on the hydrogen production via the Mn_3O_4-NaOH thermochemical cycle. *Int. J. Hydrogen Energy* **2016**, *41*, 113–122. [CrossRef]
44. Shu, S.; Guo, J.; Li, J.; Fang, N.; Li, J.; Yuan, S. Effect of post-treatment on the selective catalytic reduction of NO with NH_3 over Mn_3O_4. *Mater. Chem. Phys.* **2019**, *237*, 121845. [CrossRef]
45. Xu, X.; Li, L.; Yu, F.; Peng, H.; Fang, X.; Wang, X. Mesoporous high surface area NiO synthesized with soft templates: Remarkable for catalytic CH_4 deep oxidation. *Mol. Catal.* **2017**, *441*, 81–91. [CrossRef]
46. Gil, A.; Gandía, L.M.; Korili, S.A. Effect of the temperature of calcination on the catalytic performance of manganese- and samarium-manganese-based oxides in the complete oxidation of acetone. *Appl. Catal. A Gen.* **2004**, *274*, 229–235. [CrossRef]
47. Chen, Z.; Yang, Q.; Li, H.; Li, X.; Wang, L.; Tsang, S.C. $Cr-MnO_x$ mixed-oxide catalysts for selective catalytic reduction of NO_x with NH_3 at low temperature. *J. Catal.* **2010**, *276*, 56–65. [CrossRef]
48. Dong, Y.; Zhao, Y.; Zhang, J.-Y.; Chen, Y.; Yang, X.; Song, W.; Wei, L.; Li, W. Synergy of Mn and Ni enhanced catalytic performance for toluene combustion over Ni-doped α-MnO_2 catalysts. *Chem. Eng. J.* **2020**, *388*, 124244. [CrossRef]
49. Zhang, Y.; Qin, Z.; Wang, G.; Zhu, H.; Dong, M.; Li, S.; Wu, Z.; Li, Z.; Wu, Z.; Zhang, J.; et al. Catalytic performance of MnO_x–NiO composite oxide in lean methane combustion at low temperature. *Appl. Catal. B Environ.* **2013**, *129*, 172–181. [CrossRef]
50. Zhang, X.; Yang, Y.; Zhu, Q.; Ma, M.; Jiang, Z.; Liao, X.; He, C. Unraveling the effects of potassium incorporation routes and positions on toluene oxidation over α-MnO_2 nanorods: Based on experimental and density functional theory (DFT) studies. *J. Colloid Interface Sci.* **2021**, *598*, 324–338. [CrossRef]
51. Greluk, M.; Rotko, M.; Turczyniak-Surdacka, S. Comparison of catalytic performance and coking resistant behaviors of cobalt- and nickel based catalyst with different Co/Ce and Ni/Ce molar ratio under SRE conditions. *Appl. Catal. A Gen.* **2021**, *590*, 117334. [CrossRef]
52. Grzona, C.B.; Lick, I.D.; Rodriguez Castellón, E.; Ponzi, M.I.; Ponzi, E.N. Cobalt and KNO_3 supported on alumina catalysts for diesel soot combustion. *Mater. Chem. Phys.* **2010**, *123*, 557–562. [CrossRef]
53. Morrow, B.A.; Cody, I.A.; Moran, L.E.; Palepu, R. An infrared study of the adsorption of pyridine on platinum and nickel. *J. Catal.* **1976**, *44*, 467–476. [CrossRef]
54. Tarach, K.A.; Śrębowata, A.; Kowalewski, E.; Gołąbek, K.; Kostuch, A.; Kruczała, K.; Girman, V.; Góra-Marek, K. Nickel loaded zeolites FAU and MFI: Characterization and activity in water-phase hydrodehalogenation of TCE. *Appl. Catal. A Gen.* **2018**, *568*, 64–75. [CrossRef]
55. Verhoef, R.W.; Asscherm, M. The work function of adsorbed alkalis on metals revisited: A coverage-dependent polarizability approach. *Surf. Sci.* **1997**, *391*, 11–18. [CrossRef]
56. Ramírez de la Piscina, P.; Homs, N. Use of biofuels to produce hydrogen (reformation processes). *Chem. Soc. Rev.* **2008**, *37*, 2459–2467. [CrossRef]
57. Espinal, R.; Taboada, E.; Molins, E.; Chimentao, R.J.; Medinac, F.; Llorca, J. Cobalt hydrotalcites as catalysts for bioethanol steam reforming. The promoting effect of potassium on catalyst activity and long-term stability. *Appl. Catal. B Environ.* **2012**, *127*, 59–67. [CrossRef]
58. Miyamoto, Y.; Akiyama, M.; Nagai, M. Steam reforming of ethanol over nickel molybdenum carbides for hydrogen production. *Catal. Today* **2009**, *146*, 87–95. [CrossRef]
59. Subramani, V.; Song, C. Advances in catalysis and processes for hydrogen production from ethanol reforming. *Catalysis* **2007**, *20*, 65–106.
60. Greluk, M.; Gac, W.; Rotko, M.; Słowik, G.; Turczyniak-Surdacka, S. Co/CeO_2 and Ni/CeO_2 catalysts for ethanol steam reforming: Effect of the cobalt/nickel dispersion on catalysts properties. *J. Catal.* **2021**, *393*, 159–178. [CrossRef]
61. Biesinger, M.C.; Payne, B.P.; Grosvenor, A.P.; Lau, L.W.M.; Gerson, A.R.; Smart, R.S.C. Resolving surface chemical states in XPS analysis of first row transition metals, oxides and hydroxides: Cr, Mn, Fe, Co and Ni. *Appl. Surf. Sci.* **2011**, *257*, 2717–2730. [CrossRef]
62. Azmi, R.; Trouillet, V.; Strafela, M.; Ulrich, S.; Ehrenberg, H.; Bruns, M. Surface analytical approaches to reliably characterize lithium ion battery electrodes. *Surf. Interface Anal.* **2018**, *50*, 43–51. [CrossRef]
63. Biesinger, M.C.; Payne, B.P.; Lau, L.W.M.; Gerson, A.; Smart, R.S.C. X-ray photoelectron spectroscopic chemical state quantification of mixed nickel metal, oxide and hydroxide systems. *Surf. Interface Anal.* **2009**, *41*, 324–332. [CrossRef]
64. Bengaard, H.S.; Alstrup, I.; Chorkendorff, I.; Ullmann, S.; Rostrup-Nielsen, J.R.; Nørskov, J.K. Chemisorption of methane on Ni(100) and Ni(111) surfaces with preadsorbed potassium. *J. Catal.* **1999**, *187*, 238–244. [CrossRef]

65. Snoeck, J.-W.; Froment, G.F. Steam/CO_2 reforming of methane. Carbon formation and gasification on catalysts with various potassium contents. *Ind. Eng. Chem. Res.* **2002**, *41*, 3548–3556. [CrossRef]
66. Juan-Juan, J.; Román-Martínez, M.C.; Illán-Gómez, M.J. Effect of potassium contenet in the activity of K-promotes Ni/Al_2O_3 catalysts for the dry reforming of methane. *Appl. Catal A Gen.* **2006**, *301*, 9–15. [CrossRef]

Article

Kinetic Study and Modeling of the Degradation of Aqueous Ammonium/Ammonia Solutions by Heterogeneous Photocatalysis with TiO$_2$ in a UV-C Pilot Photoreactor

Juan C. García-Prieto [1], Luis A. González-Burciaga [2], José B. Proal-Nájera [2] and Manuel García-Roig [1,*]

[1] Centro de Investigación y Desarrollo Tecnológico del Agua, Universidad de Salamanca, Campo Charro s/n, 37080 Salamanca, Spain; jcgarcia@usal.es
[2] Instituto Politécnico Nacional, CIIDIR-Unidad Durango, Calle Sigma 119, Fracc. 20 de Noviembre II, Durango 34220, Mexico; luis.gonzalez.iq@gmail.com (L.A.G.-B.); jproal@ipn.mx (J.B.P.-N.)
* Correspondence: mgr@usal.es

Abstract: The degradation mechanism of NH_4^+/NH_3 in aqueous solutions by heterogeneous photocatalysis (TiO_2/SiO_2) and photolysis in UV-C pilot photoreactor has been studied. Under the conditions used, NH_4^+/NH_3 can be decomposed both by photolytically and photocatalytically, without disregarding stripping processes. The greatest degradation is achieved at the highest pH studied (pH 11.0) and at higher lamp irradiation power used (25 W) with degradation performances of 44.1% (photolysis) and 59.7% (photocatalysis). The experimental kinetic data fit well with a two parallel reactions mechanism. A low affinity of ammonia for adsorption and surface reaction on the photocatalytic fiber was observed (coverage not higher than 10%), indicating a low influence of surface phenomena on the reaction rate, the homogeneous phase being predominant over the heterogeneous phase. The proposed reaction mechanism was validated, confirming that it is consistent with the photocatalytic and photolytic formation of nitrogen gas, on the one hand, and the formation of nitrate, on the other hand. At the optimal conditions, the rate constants were $k_3 = 0.154$ h^{-1} for the disappearance of ammonia and $k_1 = 3.3 \pm 0.2 \; 10^{-5}$ h^{-1} and $k_2 = 1.54 \pm 0.07 \; 10^{-1}$ h^{-1} for the appearance of nitrate and nitrogen gas, respectively.

Keywords: TiO$_2$ photocatalysis; UV-C; photolysis; ammonia; nitrogen removal; water treatment

Citation: García-Prieto, J.C.; González-Burciaga, L.A.; Proal-Nájera, J.B.; García-Roig, M. Kinetic Study and Modeling of the Degradation of Aqueous Ammonium/Ammonia Solutions by Heterogeneous Photocatalysis with TiO$_2$ in a UV-C Pilot Photoreactor. *Catalysts* 2022, *12*, 352. https://doi.org/10.3390/catal12030352

Academic Editors: Magdalena Zybert and Katarzyna Antoniak-Jurak

Received: 25 January 2022
Accepted: 15 March 2022
Published: 21 March 2022

Publisher's Note: MDPI stays neutral with regard to jurisdictional claims in published maps and institutional affiliations.

Copyright: © 2022 by the authors. Licensee MDPI, Basel, Switzerland. This article is an open access article distributed under the terms and conditions of the Creative Commons Attribution (CC BY) license (https://creativecommons.org/licenses/by/4.0/).

1. Introduction

One of the major concerns regarding environmental problems is the high concentration of nitrogen in surface and groundwater from industrial discharges, urban wastewaters and poor agricultural practices leading to over-fertilization of soils that allows the diffuse contamination of nitrogen and phosphorus, among others. According to the EEA briefing 'National Emission Ceilings (NEC) Directive reporting status 2019', emissions of ammonia from the agricultural sector continue to rise. The emissions increased by 0.4% from 2016 to 2017 with a general increase of 2.5% between 2014 and 2017, posing a challenge for EU Member States in meeting EU pollution limits. The EEA briefing notes that a more substantial reduction will be required for all pollutants if the EU is to achieve its emission reduction commitments by 2030 [1].

Ammonia emissions can lead to an increased acid deposition and excessive nutrient levels in soil, rivers or lakes, which can have a negative impact on aquatic ecosystems and cause damage to forests, crops and other vegetation. Inorganic compounds are very common in urban and industrial wastewaters, as well as nitrogen compounds, ammonia and nitrates formed from the decomposition of organic nitrogen compounds [2]. The discharge of nitrogen compounds, especially NH_4^+/NH_3, into receiving watercourses causes serious problems: in aquatic ecosystems, eutrophication, reduction of dissolved oxygen, lethality in fish [3,4] and in human health, both directly, as they attack the respiratory system, skin and eyes, and indirectly, as they generate chloramines, carcinogenic compounds, in chlorination treatments to obtain tap water for consumption [5]. They also cause corrosion in copper pipes [6].

NH_4^+/NH_3 removal in industrial waters is carried out with technologies, such as chemical precipitation, electrolysis, adsorption, stripping, ultrafiltration and ion exchange [7–11]. NH_4^+/NH_3 removal in urban wastewaters is carried out either by biological treatments by heterotrophic bacteria under anoxic conditions and high COD/N ratio, or by autotrophic bacteria under anaerobic conditions (anammox processes) under low COD/N ratio conditions [12]. In low-cost wastewater treatment plants, nitrogen removal is carried out by natural phyto and geo-depuration systems [13]. However, these processes are costly and often ineffective. For example, the runoff water from urban WWTP sludge, on which this study is based, after dewatering treatment, contains concentrations of around 1000 mg/L of NH_4^+/NH_3, which cannot be discharged directly into the public watercourse and significantly increases the concentration of nitrogen to be treated in the plant, when it is recirculated to the plant headworks [14]. Traditional treatment methods are not effective enough to remove such high concentrations of NH_4^+/NH_3 [15].

Advanced oxidation process technologies and heterogeneous photocatalysis by TiO_2, in particular, have been demonstrated to be a technology capable of decomposing NH_4^+/NH_3 into harmless gases, such as nitrogen and hydrogen [16–20], proving to be an effective solution to this problem. NH_4^+/NH_3 degradation by immobilized TiO_2, as a photocatalyst, has been the most studied and widely used due to its abundance, non-toxicity, chemical stability, excellent UV light activity and also to avoid the limitation of separating TiO_2 powder from the suspension after cleaning wastewater [21]. Moreover, different immobilizations of semiconductors on solid supports have been studied [22–26]. The photocatalytic oxidation of NH_4^+/NH_3 in aqueous phase has been mainly investigated controlling solution pH, catalyst concentration [27–33] and intensity of irradiant light [16,34]. This advanced oxidation process involving photo-generated hydroxyl radicals as primary oxidants oxidizes NH_4^+/NH_3 to either nitrogen gas [35], to nitrite and nitrate anions, or both [36].

Furthermore, since 1959 it had been observed that the photolysis of ammonia occurs in the presence of nitrogen oxides to give rise to nitrogen gas [37]. Ammonia degradation by direct photolysis with hydroxyl radicals [38,39] and photo-electrochemical process [40] have been under research recently. The photochemical process based on UV sources could be a feasible technology for the treatment of NH_4^+/NH_3^- containing wastewaters.

There may be a synergistic effect between the photolytic and photocatalytic processes occurring in the photoreactors to explicate the mechanism for the removal of NH_4^+/NH_3 from aqueous solutions. This paper presents the sensitivity analysis of the main variables and the mechanism for ammonia removal by heterogeneous photocatalysis using TiO_2 supported on SiO_2 in a pilot UV-C photoreactor, considering both photolytic and photocatalytic processes.

2. Results

2.1. Kinetic Study of Ammonia Removal

Some authors have studied ammonia degradation reactions using TiO_2 photocatalyst. These authors indicate that hydroxyl ions are generated from the decomposition of water, which facilitates the oxidation of ammonia to nitrogen gas or nitrate depending on the mechanism followed [14,41,42].

A basic pH benefits the reduction of ammonia/ammonium:

$$NH_4^+ + H_2O \rightarrow NH_3 + H_3O^+ \quad pKa\,(NH_4^+) = 9.246$$

$$NH_3 + OH^\bullet \rightarrow NH_2 + H_2O + e^-$$

$$NH_2 + OH^\bullet \rightarrow NH + H_2O + e^-$$

$$N + N \rightarrow N_2$$

Some intermediate species react with each other in the presence of protons to form nitrogen gas as well:

$$NH_x + NH_y \rightarrow N_2H_{x+y} \ (x, y = 0, 1, 2) + H^+ \rightarrow N_2$$

On the other hand, a part of the ammonia species is oxidized to nitrite, either by the reaction of the oxygen with two holes created on the surface of the photocatalyst or with the solvent molecules [41]:

$$NH_3 + O_2 + 2\ \hbar^+ \rightarrow NO_2 + 3\ H^+$$

$$NH_3 + 2\ H_2O + 6\ \hbar^+ \rightarrow NO_2 + 7\ H^+$$

Nitrites, in turn, oxidize into nitrates:

$$NO_2 + H_2O + 2\ \hbar^+ \rightarrow NO_3 + 2\ H^+$$

Whether the formation of one product or the other is favored depends on the characteristics of the medium, the pH and the oxygen concentration, either favoring a basic pH, in the absence of oxygen the decomposition of ammoniacal nitrogen into gaseous nitrogen or in combination [43].

Serewicz and Noyes [37] observed that photolysis of ammonia occurs in the presence of nitrogen oxides to give rise to nitrogen gas, nitrous oxide and small amounts of hydrogen. Other authors [38,39] have investigated the oxidation pathway of photolytic ammonia oxidation in presence of $OH^{\bullet-}$ radical. Huang et al. [38], using the laser flash photolysis technique with transient absorption spectra of nanosecond, proposes that hydroxyl ions could oxidize NH_3 to form NH_2 as the main product of photolytic oxidation. It would further react with OH$^-$ radical to yield NHOH. Since NHOH could not remain stable in solution, it would rapidly convert to $NH_2O_2^-$ and consequently NO_2^- and NO_3^-. Wang et al. [39] suggest the following reaction mechanism:

$$NH_3 + 6\ OH^{\bullet} \rightarrow NO_2 + 4\ H_2O + H^+$$

$$NO_2 + 2\ OH^{\bullet} \rightarrow NO_3 + H_2O \ (rapid)$$

$$NO_3 + H_2O + h\nu \rightarrow NO_2 + O_2 + 2\ H^+ \ (slow)$$

In order to study the optimal conditions for ammonia degradation, as well as the contribution of the photocatalytic and photolytic processes, a series of studies were carried out, as follows.

2.2. Influence of Different Factors on the Heterogenous Photocatalytic Process of NH_4^+/NH_3 Removal

Firstly, a sensitivity analysis of the optimal conditions for NH_4^+/NH_3 removal by heterogeneous photocatalysis in the UBE photoreactor was carried out. It is known that the factors that most influence NH_4^+/NH_3 degradation the most, in addition to temperature, are pH, conductivity, dissolved oxygen and lamp irradiation power (P).

The influence of pH is due to the fact that the photocatalytic reactions take place on the surface of the photocatalyst and they are strongly influenced by that surface charge, which is different in acidic or alkaline conditions. The TiO_2 isoelectric point is around 6.5 [44], then at lower solution pH its surface would be positively charged, whereas at higher pH the surface would be negatively charged [45]. Consequently, the electrostatic properties of catalysts' surfaces in different environments with respect to reactive compounds play an important role on the rate of the reaction [46]. Furthermore, the speciation of the NH_4^+/NH_3 acid/conjugate base pair at the different working pHs must be taken into account. pKa of ammonium is 9.25, thus, at pH 7.0, TiO_2 is near to neutrality and NH_4^+ (99.4%) has a positive charge and no electrostatic attractive forces appear, at pH 9.0 TiO_2 is negatively charged and NH_3 (36%) / NH_4^+ (64%), then significant attractive forces between photocatalyst surfaces and NH_4^+ occur, finally at pH 11.0 TiO_2 is negatively charged and NH_3 (98.3%) /NH_4^+ (1.7%) and almost no attractive forces should appear. Considering that degradation of NH_4^+/NH_3 increases at higher pHs, the authors infer that attractive forces between pollutant particles and photocatalyst surface do not play an essential role in these experimental conditions. Several authors indicate that ammonia oxidation takes place only under alkaline conditions [27,29,34,47]. In this sense, others authors proposed that at basic pH, the scenario is different and the adsorption of neutral NH_3, rather than NH_4^+ on the surface of TiO_2, may be the rate-limiting step [28]. Thus, Shibuya et al. [32] observed at pH 10 that increasing the amount of molecular ammonia adsorbed by the catalyst is important for enhancing the oxidation rate. At the same time, at high pHs, the negative charge of the photocatalyst surface hinders the adsorption of other anions and aromatic organic compounds as they probably exist as anionic species at these pHs [48–50]. Furthermore, at higher pHs the concentration of H^+ significantly decreases, reducing its competition with NH_4^+ for exchange sites on the photocatalyst surface, thus favoring the efficiency of ammonium removal [16,27,31].

Figure 1 shows the kinetic curves of the concentrations of ammonium/ammonia, nitrate and nitrogen gas. The kinetic curve for the concentration of nitrogen gas released is calculated by the difference after the corresponding mass balance. Under these conditions the values for gas-phase species are relative values, not absolute values.

The nitrite concentration measured was always in the order of 100 times lower than the nitrate concentration, but since the rate of nitrite oxidation at lower pH is faster than at a higher pH [32], then the nitrite concentration should increase with pH. Thus, it is considered that nitrite is a very unstable intermediate species that is rapidly transformed into nitrate or nitrogen gas and therefore the nitrite concentration has not been considered, nor represented, for kinetic purposes. In this respect, Wang et al. [39] recently observed that ammonia is more reactive than ammonium ion with hydroxyl radical (OH.), by a stepwise H_2O_2 addition method, NH_4^+/NH_3 can be completely converted to NO_x^- in UV-C photolytic process. The low formation of nitrate could be explained by pH effect and another route involves photo-reduction of nitrate to form nitrite or nitrogen gas by photolytic and photocatalytic processes [40,51–53].

Other authors [31,42] indicate that when the concentration of dissolved oxygen increases significantly, the concentration of nitrite and nitrate in the solution increases as well. In the experiments carried out in this work, no oxygen has been introduced into the solution, except for the oxygenation that may occur as a consequence of the recirculation flow rate (1000 L/h) of the sample in the photoreactor circuit. At the experimental conditions followed in the NH_4^+/NH_3 degradation kinetics, measurements of the dissolved oxygen concentration of the samples have shown that there are no noticeable changes in the dissolved oxygen concentration, as shown in Figure 2.

However, conductivity of the samples increases with time in the ammonium degradation experiments, observing that the difference between the initial conductivity (time zero) and the final conductivity (7 h) is greater with increasing pH and lamp irradiation power, which indicates that more ionic intermediate species are being formed (Figure 3).

Therefore, after this sensitivity analysis, it could be stated that the percentage of NH_4^+/NH_3 removal is mainly a function of lamp irradiation power and pH. In this sense, the results shown in Figure 4 indicate that the degree of NH_4^+/NH_3 removal increases with pH at both lamp irradiation powers. This is consistent with what has been observed by other authors for homogeneous photocatalytic processes [16,31]. Considering that pKa of ammonium is 9.25, the ammonium ion is the major species at pH 7.0 and pH 9.0, while molecular ammonia is predominant at pH 11.0.

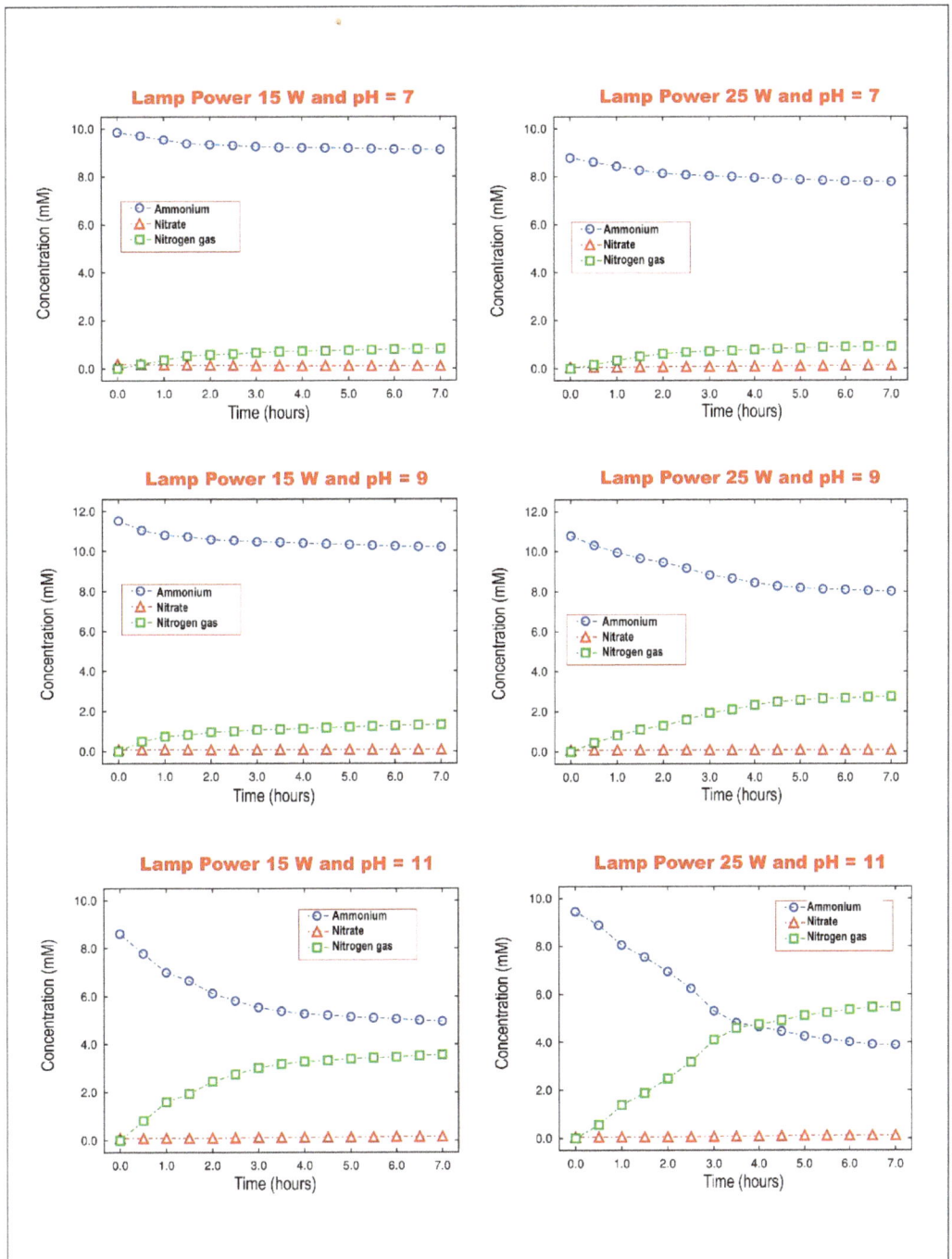

Figure 1. Influence of pH (pH 7.0–11.0) and lamp irradiation power (15 W, 25 W) on the performance and kinetics of NH_4^+/NH_3 removal by photocatalysis. Q = 1000 L/h, temperature 20 ± 1 °C.

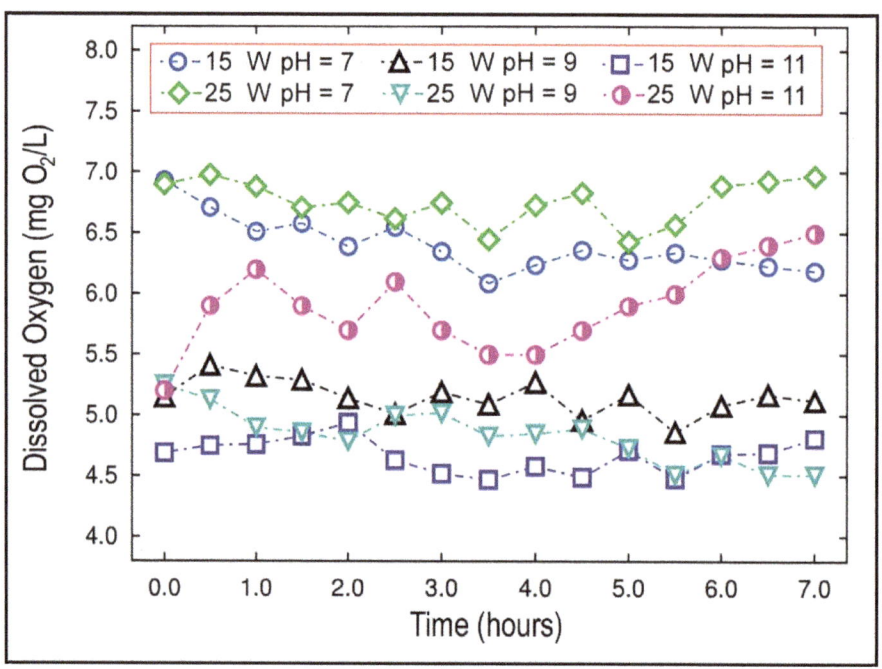

Figure 2. Dynamics of dissolved oxygen in the samples during NH_4^+/NH_3 removal process by photocatalysis. Q = 1000 L/h, temperature = 20 ± 1 °C.

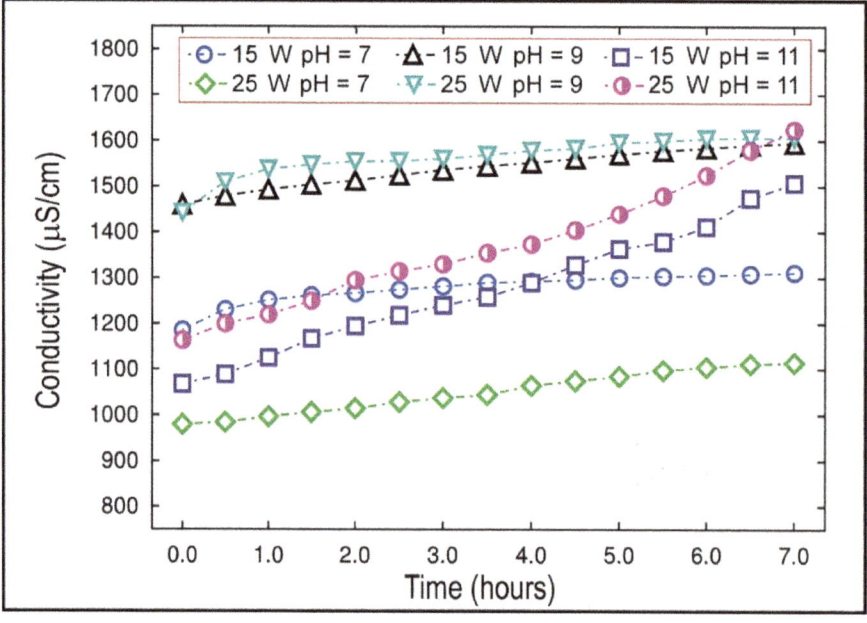

Figure 3. Dynamics of sample conductivity during NH_4^+/NH_3 removal process by photocatalysis. Q = 1000 L/h, temperature = 20 ± 1 °C.

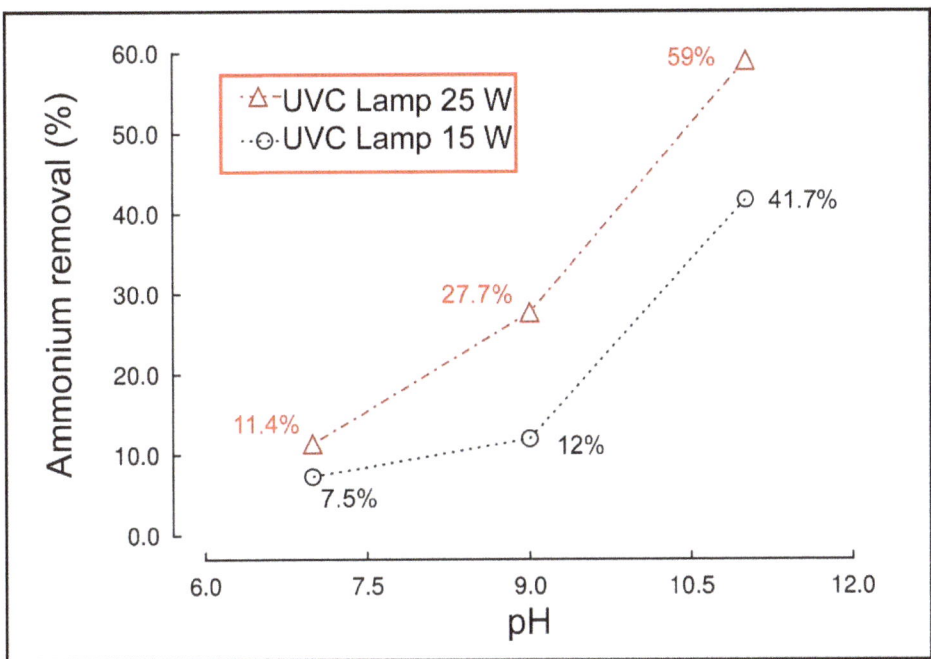

Figure 4. Dependence of NH_4^+/NH_3 removal performance (at 7 h), by photocatalysis, on pH (7.0–11.0) and lamp irradiation power (15 W, 25 W). Q = 1000 L/h, temperature = 20 ± 1 °C.

Previous studies of fiber stability against acids and bases indicates that the fiber is stable with respect to pH, except at extreme pHs, with loss of titanium at very acidic pHs and loss of silicon at very basic pHs (higher pH = 11) [54].

2.3. Influence of Stripping, Photocatalysis and Photolysis on the Kinetics of Ammonia Removal

According to the optimal conditions, three experiments were carried out with similar initial ammonium concentrations (180–200 mg/L), at pH = 11.0, 25 W UV-C lamp, Q = 1000 L/h and T = 20 ± 1 °C (Figure 5). This was done so in order to study the effect of ultraviolet light (photolysis effect) and the effect of stripping or degassing, as a consequence of the recirculation flow rate of the sample (Q = 1000 L/h), on the performance of ammonia removal by heterogeneous photocatalysis. This stripping will produce a decrease in the concentration of ammonium in the solution as a result of the passage, by volatilization, of the dissolved ammonium into the air. In this sense, a series of experiments were carried out as follows in order to check the different effects on the kinetics of ammonia degradation: the first experiment with the lamp on and photocatalyst (photocatalytic effect), the second experiment with the lamp off and photocatalyst (stripping effect) and the final experiment without photocatalyst but with the lamp on (photolysis effect). Figure 5 shows the percentage of degradation as a function of reaction time in each of the three experiments carried out. The highest performance is by photocatalysis (59.7 %) but a high removal by photolysis (44.1%) is also observed, which may be caused by photochemical reactions (photolysis) with nitrogen oxides formed during recirculation (stripping) with ammonia in the presence of UV-C light [37,40]. In this context, the effect of nitrogen removal by stripping (14.4%) is significant, which suggests that part of the ammonia may be lost from the solution either by volatilization, adsorption on the catalyst surface or both, not to mention the fact that ammonia may oxidize into nitrogen oxides due to agitation and aeration of the mixture during recirculation of the solution.

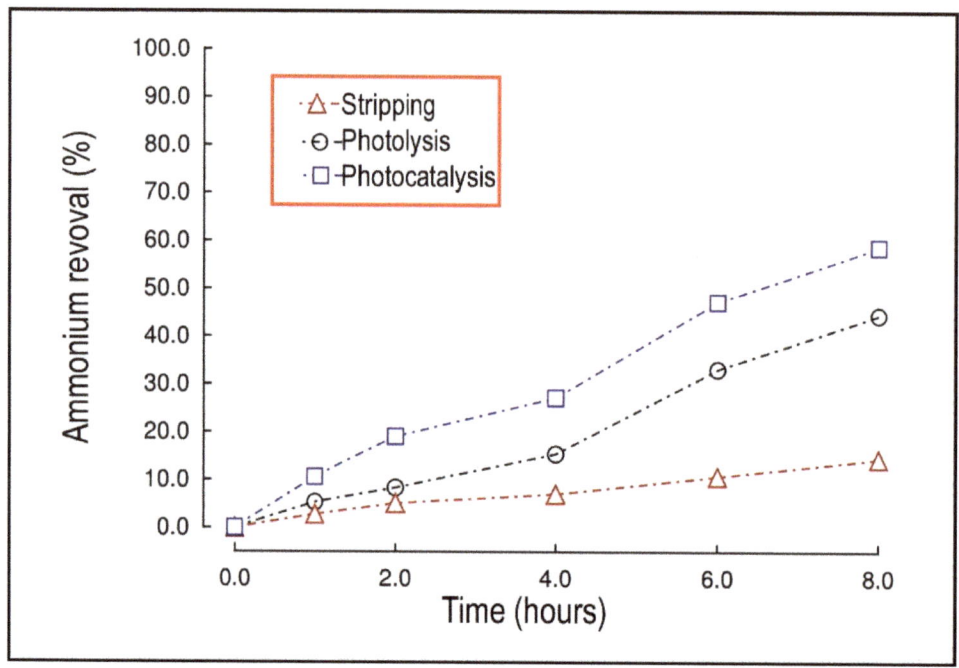

Figure 5. Percentage of ammonia removed vs. time by photolysis (o), photocatalysis (□) and stripping (Δ) treatments at pH = 11.0, 25 W UV-C lamp. Q = 1000 L/h, temperature = 20 ± 1 °C.

The time range used in this work and the yields achieved are similar to the processes currently used to remove ammonia nitrogen, such as the annamox process (8-h cycles), responsible for 50% of the nitrogen turnover in marine environments at various temperature and salinity conditions [55].

Considering that the stripping effect is very similar in all the experiments performed at the same recirculation flow rate, it can be considered that the ammonia removal would be carried out by competitive or parallel reactions, following kinetics of order one. Expressing the kinetic equations of the model in an integrated way:

$$C = C_o \, e^{-(k_{PC}+k_{PL})t}$$

$$C = C_o \, e^{-k_{op} \cdot t}$$

where C and C_o are the ammonia concentrations at times t and 0, respectively; k_{PC} is the rate constant of the photocatalytic process and k_{PL} is the rate constant of the photolysis process, the operational constant k_{op} being equal to the sum of both constants. The photocatalytic rate constant (k_{PC}) would in turn be an apparent constant that would encompass the reactions of ammonia oxidation to nitrogen gas and nitrate depending on a reaction mechanism that is difficult to elucidate.

Figure 6 shows the operational rate constants (k_{op}) for NH_4^+/NH_3 removal at different pHs and lamp irradiation powers, following a kinetic curve of monoexponential progress to the baseline. According to the observation, the degradation does not seem to tend to zero, i.e., total elimination of the compound, but to an asymptotic degradation value (C). The results were fitted to a monoexponential model with decrease to a baseline (y = a e^{-bx} + C, where C is the saturation value). This fact has also been observed by Murgia et al. [42].

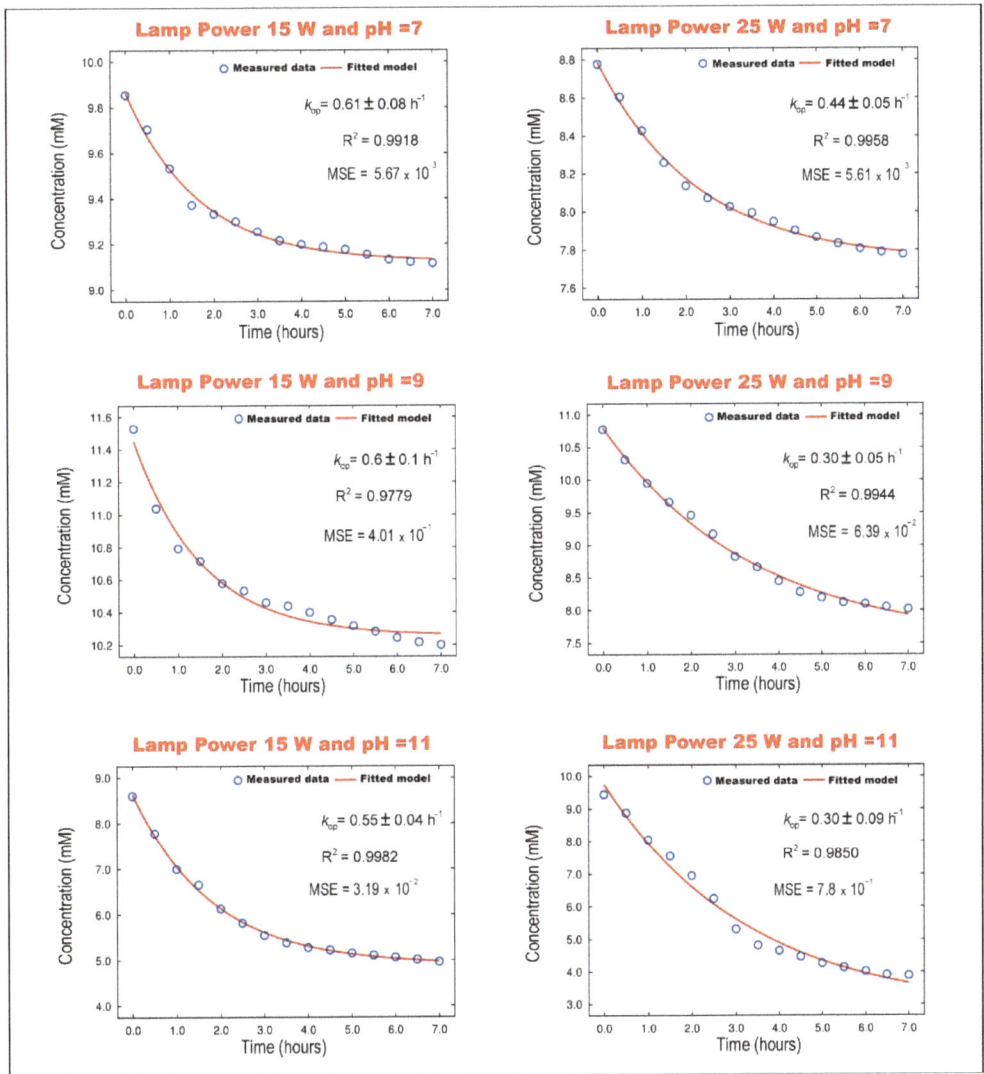

Figure 6. Fit of the kinetics of NH_4^+/NH_3 degradation by photocatalysis to a monoexponential progress kinetic curve to baseline.

The fits are good (coefficient of determination R^2 = 0.9779–0.9982, mean squared error MSE = 5.61×10^{-3}–7.8×10^{-1}) but there is no agreement between the degree of NH_4^+/NH_3 removal and the operational rate constants obtained at different pHs and lamp irradiation powers, since the operational rate constant (k_{op}), according to the observed experimental facts, should increase with pH and lamp irradiation power. Thus, from the results obtained, it could be thought that in the NH_4^+/NH_3 removal, intermediate species formed could either participate in other types of reactions or that different species of the reaction mechanism are involved in different reactions on the surface of the photoreactor or in combination, being adsorbed and saturating the photoreactor surface (asymptotic value).

In this sense, it could also be thought that NH_4^+/NH_3 species are not adsorbed, intermediate species are, which would explain why the removal values tend towards an asymptotic value and the degradation rate does not increase, according to the fits regarding pH and irradiation power of the lamp (Figure 6), contrary to what was expected. On the one hand, the increase in NH_4^+/NH_3 removal has a strong effect on the ammonia species (NH_3) but not so much on the ammonium ion (NH_4^+), which could indicate that the two species behave differently on the catalyst surface as asserted by some authors [16,28,32]. On the other hand, the increased conductivity and low formation of nitrite seems to indicate that this species is rapidly formed and decomposed. Nitrite can participate in both the photocatalytic oxidation of nitrite into nitrate [42], and in the photolysis decomposition of intermediate ammonium species (NH_x) to nitrogen gas [37]. Finally, it should be noted that, according to the mass balance carried out, it is observed that the formation of nitrogen gas always reaches a maximum, i.e., saturation of the photocatalyst surface occurs, which may be due to intermediate species that saturate the active sites of the fiber surface, causing a decrease in the saturation rate that corresponds to the values of the operational rate constants shown in Figure 6.

It is interesting to note that the experimental data were also fitted (data not shown) according to the model equations for the case of NH_4^+/NH_3 removal following a monoexponential progress kinetic curve with a tendency to zero, i.e., total NH_4^+/NH_3 elimination. However, in this case the results show a poor fit of this model to the experimental data, except for the case of the largest performance of the reaction at pH 11.0 and 25 W irradiation power of the UV-C lamp, which gives a good fit, which would indicate that this phenomenon of reaction inhibition by adsorption of species on the surface of the photocatalyst has a greater effect the slower the rate of the reaction, i.e., at pH 7.0 and 9.0 and 15 W UV-C lamp irradiation power. Therefore, a study of the proposed mechanism under the most favorable ammonium removal conditions (pH = 11.0 and P = 25 W UVC) is addressed in the following section.

2.4. Initial Ammonia Removal Rates at pH = 11.0 and 25 W UVC Ultraviolet Lamp Irradiation Power. Models Discrimination

To discriminate between the possible existence of parallel reactions and the probable intervention of different intermediate species in the reaction mechanism, the kinetics of the degradation of ammonia is also studied by the differential method of the initial rates, measuring these as a function of the initial concentration of the substrate, reducing, under these conditions, to a minimum, the influence of such reactions and the adsorption effects on the photocatalyst surface by reactants and intermediate species to a minimum.

The best fits of the experimental kinetic curves were to the monoexponential models ($R^2 = 0.9756$–0.9947, MSE = 6.1×10^{-3}–6.0×10^{-2}) and the corresponding values of the initial rates v_o are shown on Figure 7 for the different ammonia initial concentrations. The green dotted line marks the tangent to the curve whose slope represents $-v_o$ and the black dashed line indicates the asymptotic value.

Different authors [16,18,42] state that the kinetics of photocatalytic ammonium/ammonia removal reactions follow the recommended Langmuir–Hinshelwood model, shown by the differential equation:

$$v_o = -\frac{dC}{dt} = \frac{k\,K\,C_o}{1 + K\,C_o}$$

where k is the reaction rate constant (when $1 << KCo$), K is the ammonium/ammonia adsorption equilibrium constant on the photocatalyst, C_o is the initial ammonium/ammonia concentration. The results of the fit v_o (C_o) are shown in Figure 8, where a small curvature of the line fitted to the experimental points can be seen, which together with the statistical quality of the fit ($R^2 = 0.9942$, MSE = 1×10^{-2}), seems to confirm the trend towards the behavior of the Langmuir–Hinshelwood equation and mechanism in this ammonium/ammonia concentration range.

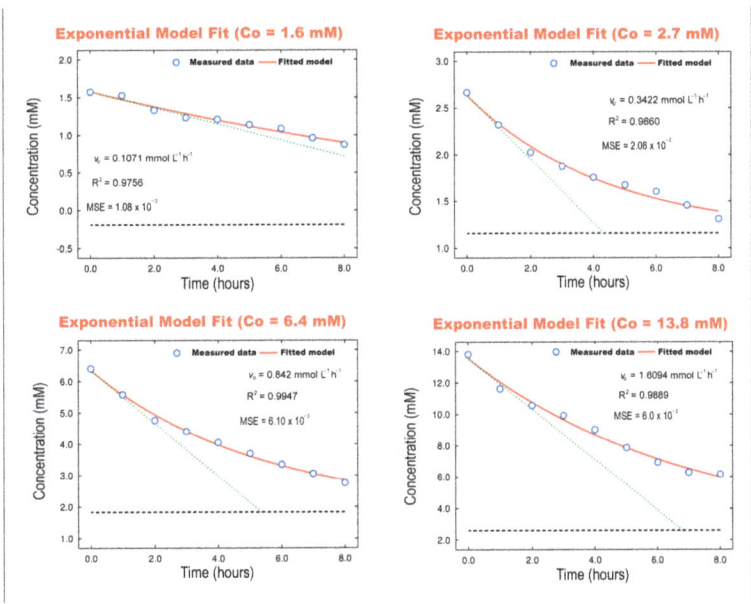

Figure 7. Initial ammonia removal rates by photocatalytic process at different initial concentrations of ammonia (at pH = 11.0, P = 25 W, Q = 1000 L/h, temperature = 20 ± 1 °C).

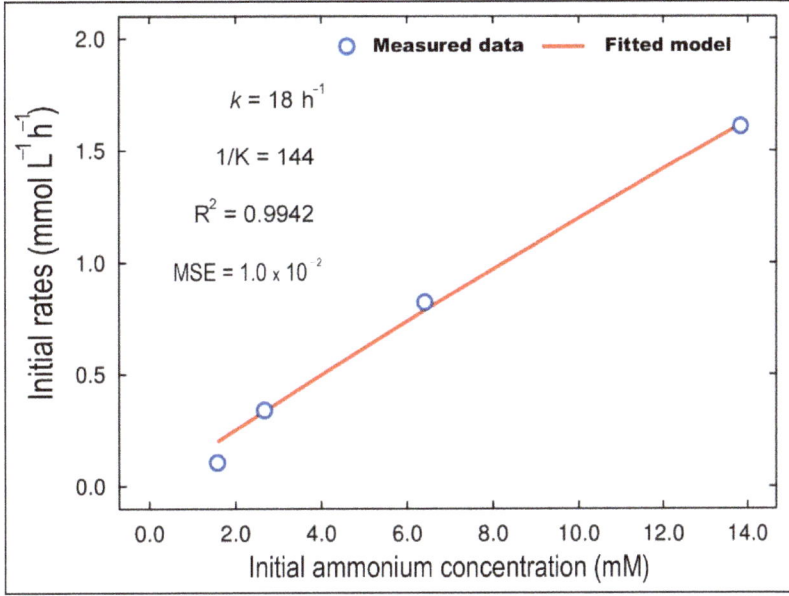

Figure 8. Fit to the Langmuir–Hinshelwood differential equation of the initial rate versus initial concentration data for ammonia removal photocatalytic process at pH = 11.0, P = 25 W, Q = 1000 L/h, temperature = 20 ± 1 °C.

Integrating the Langmuir–Hinshelwood differential equation, the corresponding integrated equation can be expressed as:

$$\frac{1}{K}\ln\left(\frac{C_t}{C_o}\right) + (C_t - C_o) + kt = 0$$

The constants obtained from Figure 8, the rate constant (k = 18 h^{-1}) and the adsorption equilibrium constant (K = 0.0069), as well as the operational constant defined as the product of both constants ($k_{op} = k\,K$ = 0.13 h^{-1}), were taken as initial estimates with which we proceeded to close the fitting of the integrated Langmuir–Hinshelwood equation for each of the global ammonium removal kinetic curves shown in Figure 7. This procedure will allow discovering of the sensitivity and accuracy of the fit over time and not only at initial times of ammonium removal reaction, thus indicating if any interference has occurred during the course of the reaction time.

Figure 9 shows the good nonlinear regression fits to the experimental points C_t (t) of the integrated rate equation of the Langmuir–Hinshelwood model (R^2 = 0.9650–0.9871, MSE = 1.1 × 10^{-2}–7.4 × 10^{-2}), as well as the values of the parameters, rate constant (k) and inverse of the equilibrium adsorption constant (1/K). A slight increase in the rate constant of the reaction is observed as a function of the concentration (k = 12–22 h^{-1}), while the value of the adsorption equilibrium constant remains constant (K = 0.005) as the ammonia initial concentration increases.

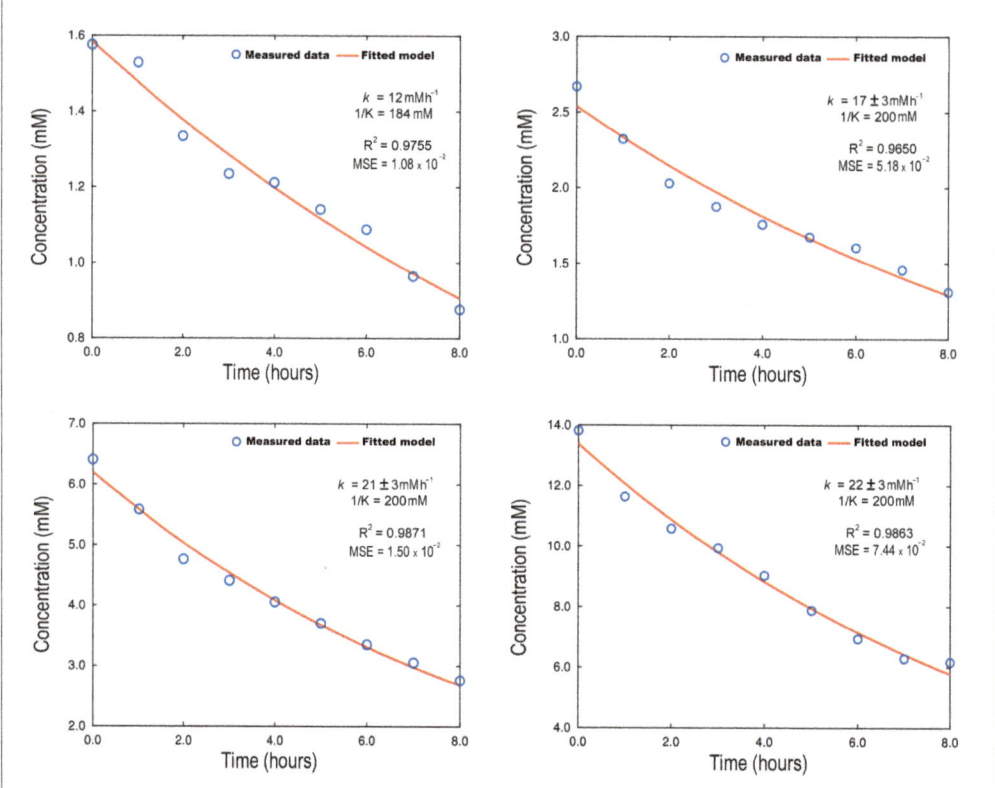

Figure 9. Fit to the integrated Langmuir–Hinshelwood rate equation of the experimental data C_t (t) of the ammonia removal photocatalytic process at different initial ammonia concentrations (pH = 11.0, P = 25 W, Q = 1000 L/h, temperature = 20 ± 1 °C).

Under the conditions studied, it is observed that the operational rate constant was very low (0.07–0.11 h^{-1}) and no significant variation of the operational rate constant with substrate concentration was observed, remaining practically constant. This indicates a low influence of surface phenomena on the reaction rate, in which the limiting step would be the transfer of matter to the surface of the photocatalytic TiO$_2$-SiO$_2$ glass fiber.

To verify this fact, the surface coverage is determined. The surface coverage Θ can be related to the substrate concentration C and the apparent adsorption constant at equilibrium K, by means of the equation:

$$\Theta = \frac{KC}{1 + KC}$$

Thus, the value of surface coverage will be between 0 and 1, as it indicates the ratio of occupied sites to total sites, tending to zero when all sites are free and to one when all sites are occupied. Table 1 shows the surface coverage values for the different ammonium initial concentrations:

Table 1. Photocatalyst surface coverage as a function of initial ammonia concentrations.

Initial Concentration (mM)	1.57	2.67	6.41	13.82
Surface coverage Θ	0.008	0.013	0.032	0.07

The surface coverage values are very low, not exceeding, at the highest concentration, 10% of the surface coating, which indicates a low adsorption of ammonia species on the surface of the photocatalytic fiber. Likewise, the operational rate constant is very low because the adsorption equilibrium constant is very low as well, which seems to indicate, according to the values obtained from the surface coverage, little adsorption of the substrate on the photocatalyst surface. This could be either because other species or intermediates are adsorbed, or because of the low affinity of this species for the active sites, in this case the homogeneous phase being predominant over the heterogeneous phase. By way of comparison, it is interesting to note that the values of the different parameters of the Langmuir–Hinshelwood equation obtained (Figure 9) are similar to those found in other studies carried out with titanium dioxide in homogeneous aqueous solutions [42]. Other authors indicate discrepancies between the degree of correlation attainable between parameters deduced from such adsorption studies (Langmuir–Hinshelwood model) and those deduced from measurements of relative efficiencies and solute concentration dependence of titanium dioxide (TiO$_2$)-sensitized photocatalyzed degradation of the model pollutants, because this should not be ignored, for poorly adsorbing pollutants, of the roles of solvent molecules at the micro-interfaces since, in reality, polar solvent molecules are likely to compete strongly against solute species for adsorption sites [56].

This assessment also seems to be in line with the high NH$_4^+$/NH$_3$ degradation performance observed by photolysis (44.1%) compared to that observed by photocatalysis (59.7%). Likewise, it seems to indicate that the reaction intermediates in the homogeneous phase react with the ammonia, mostly converting it into nitrogen gas, since the amount of NO$_2^-$ and NO$_3^-$ products formed is very low. It could also explain why the low concentration of nitrite found would favor its transformation into nitrogen gas by photolytic processes [37], which would also favor the low nitrite to nitrate transformation discovered. Even so, photocatalysis is more efficient at removing NH$_4^+$/NH$_3$ than photolysis.

It is possible that the kinetics of ammonia removal actually tends to zero, considering a monoexponential model with a tendency to zero, but it is also possible that other species saturate the photocatalytic fiber, preventing this tendency. This approximation seems to be confirmed by the C values tending to zero observed in the asymptotes (black dashed line) of Figure 7 (initial ammonia removal rates at different initial concentrations). Therefore, the progress kinetic curves $C(t)$ of the four initial rates are fitted now to monoexponential models with zero baseline trend (Figure 10), being the observed parameters consistent with the values of the operational rate constants obtained by fitting the integrated Langmuir–

Hinshelwood equation (Figure 9). This fit indicates that for the conditions set (pH 11.0), ammonium concentration tends to zero but not under experimental conditions of lower reaction rates (pH 7.0 and pH 9.0 and P 15 W). The surface of the photocatalyst can become saturated by other species that are adsorbed by the photocatalytic fiber, and then the amount of ammonia removal would tend towards an asymptotic value.

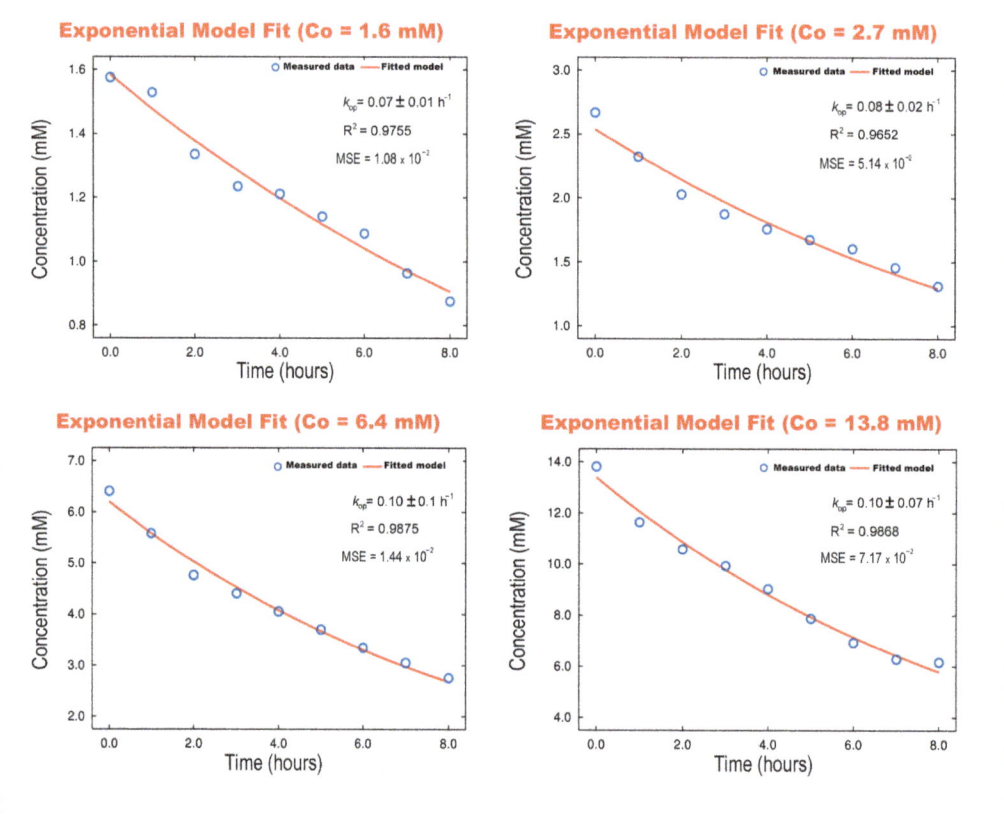

Figure 10. Fit to the monoexponential rate equation tending to zero baseline of the experimental points C_t (t) of the ammonia removal photocatalytic process at different initial ammonia concentrations (pH = 11.0, P = 25 W, Q = 1000 L/h, temperature = 20 ± 1 °C).

Finally, for comparative purposes with regard to the photocatalytic process, the photolysis process of ammonia removal is presented under the same optimal experimental conditions, that is, pH = 11.0 and P = 25 W but without photocatalytic fiber. Figure 11 shows the fit of the experimental data C (t) to the integrated rate equation of order 1, being k_{PL}, the rate constant of the photolysis process:

$$C = C_o\, e^{-k_{PL}\cdot t}$$

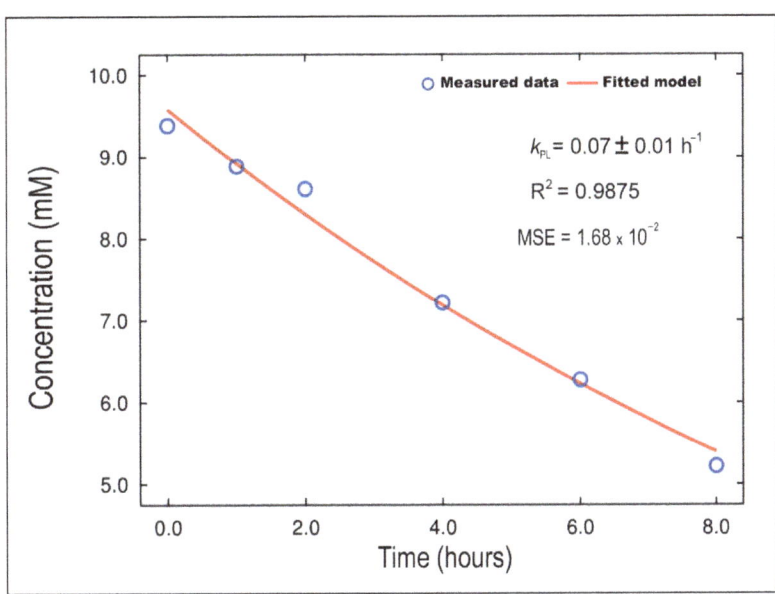

Figure 11. Fit of the experimental data $C(t)$ of the ammonia removal photolysis process to the first order monoexponential rate equation tending to zero (pH = 11.0, P = 25 W, Q = 1000 L/h, temperature = 20 ± 1 °C).

As can be seen in Figure 11, the value of the photolysis rate constant (k_{PL}) = 0.07 ± 0.01 h^{-1} is of a similar order of magnitude to the operational photocatalytic rate constant (k_{op}) of Figure 10, being greater the photocatalytic effect the higher the initial concentration of ammonia.

2.5. Study of the Photocatalytic Fiber Stability after Ammonium/Ammonia Removal Process

A study of photocatalytic fiber degradability was carried out after the NH_4^+/NH_3 degradation experiments. For this, a structural characterization study was performed, semi-quantitatively identifying, at the surface level, both the chemical elements deposited and those that form the photocatalytic fiber, as well as carrying out 50 and 200 μm scanning electron micrographs (Figure 12).

Photocatalytic fiber at pHs higher than 11.0 deteriorates by losing mass, mainly in the form of silicon, thus no studies of NH_4^+/NH_3 degradation are carried out at pH greater than 11.0. The results of Figure 12 show a coating of the fiber with impurities of chlorine (from the reagent ammonium chloride), iron, potassium and sodium deposited mainly on the titanium particles, which shows adsorption on the surface of the photocatalyst of species other than ammonium, supporting the empirical results presented above.

2.6. A Two Parallel Reaction Mechanism for Ammonium Degradation by Photolysis and Photocatalysis

According to previous studies performed by other authors [14,36–39,42] and the empirical results shown in this work, the degradation of NH_4^+/NH_3 to two final reaction products, nitrogen gas and nitrate, is carried out through two parallel reactions by simultaneous photolysis and photocatalytic actions, proposing the mechanism of Figure 13, where nitrogen gas is formed by photocatalysis and photolysis processes and nitrate is formed via photocatalysis.

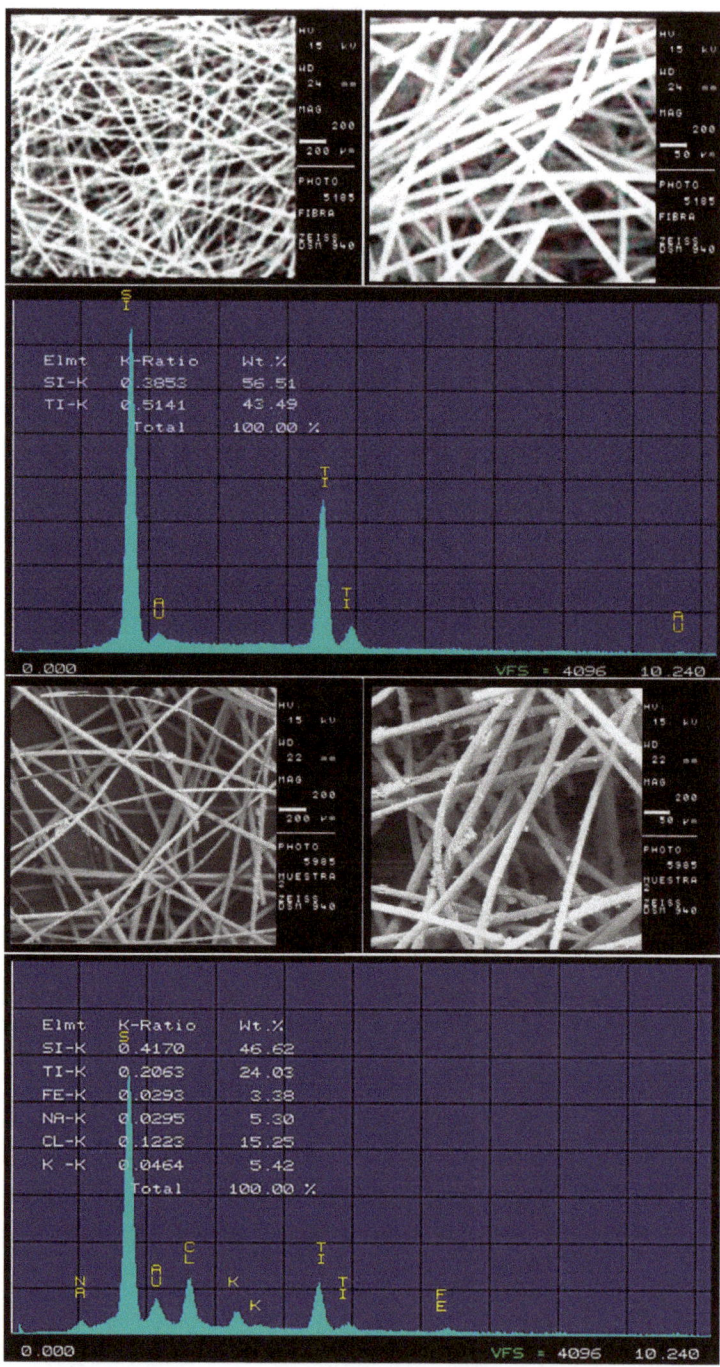

Figure 12. Study of the photocatalytic fiber stability after NH_4^+/NH_3 removal process: 200 and 50 µm scanning electron micrographs and energy dispersive X-ray elemental microanalysis (EDXMA). Upper panel: fiber virgin and lower panel: fiber after NH_4^+/NH_3 removal.

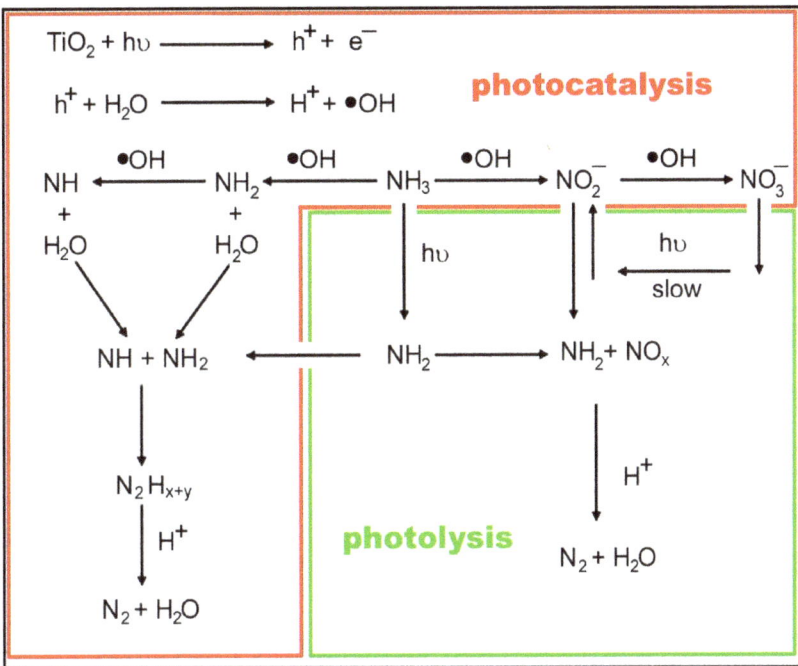

Figure 13. The proposed two parallel reaction mechanism for NH_4^+/NH_3 degradation by photolysis and photocatalysis.

In order to perform the validation of the proposed model of parallel reactions, a final confirmation of the model consisted of simultaneously carrying out the fits of the respective $C(t)$ data to the differential rate equations for the ammonium degradation and the formation of nitrate and nitrogen gas. Intermediate species, such as nitrite, have not been considered, as their formation and disappearance are very fast and the measured experimental values were found to be too low. According to the scheme presented in Figure 13, their system of differential rate equations would be:

$$v = -\frac{d[Ammonium]}{dt} = k_1[Ammonium] + k_2[Ammonium] = (k_1 + k_2)[Ammonium]$$

$$v = \frac{d[nitrate]}{dt} = k_1[Ammonium]$$

$$v = \frac{d[nitrogen\ gas]}{dt} = k_2[Ammonium]$$

where k_1 corresponds to the nitrate formation rate constant and k_2 to the nitrogen gas formation rate constant (both by photolysis and photocatalysis) and k_3 expressed as the sum of the nitrogen gas and nitrate formation rates ($k_1 + k_2$) would correspond to the ammonium/ammonia degradation overall rate constant (k_3). The results of the simultaneous fits of $C(t)$ data to the three differential equations are shown in Figure 14, the blue line showing the fit of the ammonium/ammonia degradation data, the red line, the fit of nitrate formation data and the green line, the fit of nitrogen gas formation data to the corresponding differential rate equations. Table 2 shows the values of k_1, k_2 and k_3 obtained from such fittings.

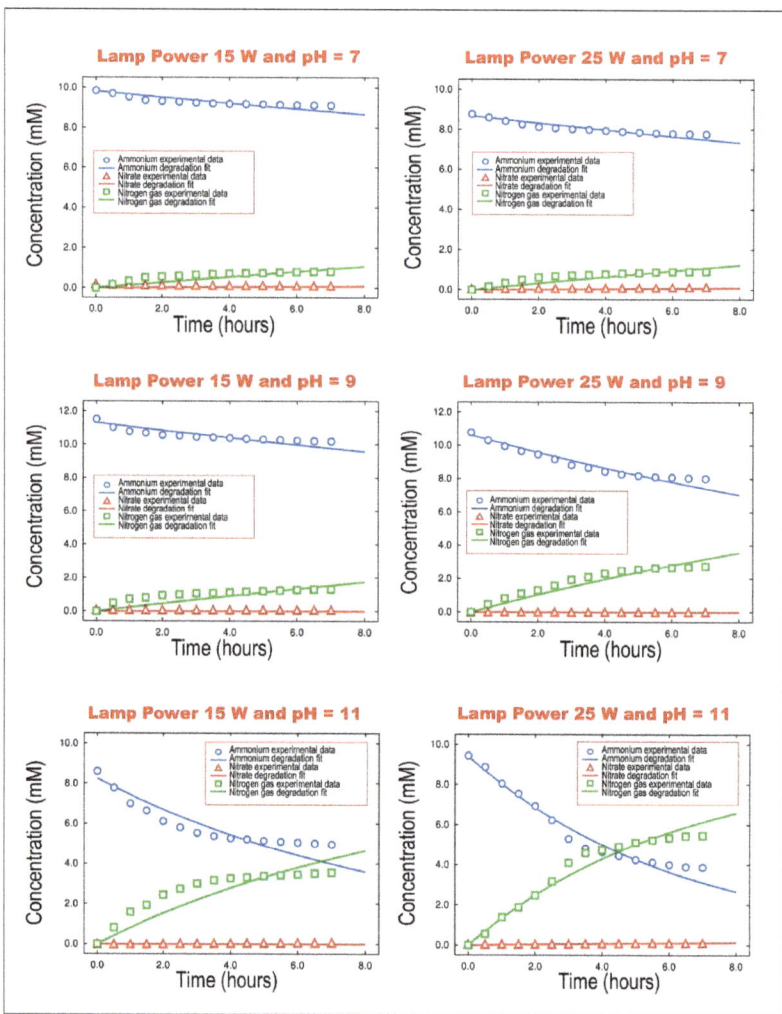

Figure 14. Simultaneous fits of 3 sets of $C(t)$ data to the corresponding 3 differential rate equations according to the proposed two parallel reaction mechanism for NH_4^+/NH_3 degradation by photolysis and photocatalysis, at two lamp irradiation powers and different pHs, Q = 1000 L/h, temperature = 20 ± 1 °C.

Table 2. Values of the rate constants for the disappearance of ammonium/ammonia (k_3) and the appearance of nitrate (k_1) and nitrogen gas (k_2), obtained by simultaneous fits of the three differential equations proposed to the corresponding sets of experimental $C(t)$ data.

	k_1 (h^{-1})	k_2 (h^{-1})	k_3 (h^{-1})
Photocatalysis 15 W pH = 7.0	$(1.1 \pm 0.2) \, 10^{-3}$	$(1.5 \pm 0.2) \, 10^{-2}$	0.016
Photocatalysis 15 W pH = 9.0	$(8.2 \pm 0.6) \, 10^{-6}$	$(2.1 \pm 0.2) \, 10^{-2}$	0.021
Photocatalysis 15 W pH = 11.0	$(2.2 \pm 0.1) \, 10^{-5}$	$(1.03 \pm 0.09) \, 10^{-1}$	0.103
Photocatalysis 25 W pH = 7.0	$(1.5 \pm 0.2) \, 10^{-3}$	$(1.9 \pm 0.5) \, 10^{-2}$	0.021
Photocatalysis 25 W pH = 9.0	$(8.7 \pm 0.6) \, 10^{-7}$	$(5.1 \pm 0.3) \, 10^{-2}$	0.051
Photocatalysis 25 W pH = 11.0	$(3.3 \pm 0.2) \, 10^{-5}$	$(1.54 \pm 0.07) \, 10^{-1}$	0.154

The simultaneous fit of the two parallel reactions model to the experimental data was good (in Student's t-test, "p-values" of the parameter estimates were less than 0.05), which confirms that the proposed model (Figure 13) fits well the experimental results. It is also observed that the degradation of NH_4^+/NH_3 into nitrogen gas is faster than the degradation of NH_4^+/NH_3 into nitrate, as would be expected from the observation of the progress curves of Figure 1. Furthermore, the values of the NH_4^+/NH_3 degradation rate constants (k_3) are in agreement with the values obtained for the operational rate constant from the fit to the integrated Langmuir–Hinshelwood equation at different initial ammonia concentrations (Figure 9) and the values of the operational rate constants obtained by fitting the experimental data to the monoexponential models with zero baseline (Figure 10). Finally, the values of the NH_4^+/NH_3 degradation constant (k_3) at the different pHs and lamp irradiation powers studied are consistent with the empirical values obtained for the NH_4^+/NH_3 removal (Figure 4), with its removal performance being higher at a higher reaction rate. All these evidences seem to confirm that the mechanism of two parallel equations fits the experimental $C(t)$ data for NH_4^+/NH_3 degradation by photolysis and heterogeneous photocatalysis.

3. Materials and Methods

3.1. TiO_2/SiO_2 Fixed Bed Photoreactor with Total Recirculation

The TiO_2/SiO_2 fixed bed photoreactor (UBE Industries, Japan) used is shown in Figure 15. The system has a tank for the sample with a capacity of 200 L, a 1 hp pump for recirculation of the sample through the system and, at the outlet of the pump, the water first goes through a 50 µm solid filter and then through a rotameter to measure the sample flow entering the photoreactor body.

The photoreactor is characterized by maintaining a vertical piston flow with bottom inlet and upper lateral outlet; the reactor body is made of stainless steel and has a manometer on the bottom and another on the top. The internal walls of the photoreactor are polished, in such way that the radiation that reaches them is reflected, thus generating greater incidence of light on the photocatalyst.

For the photocatalysis experiments, in addition to the pilot photoreactor described, a TiO_2/SiO_2 photocatalyst (UBE Industries, Japan) was used. The semiconductor material used as a non-woven photocatalytic fiber with gradient in the crystalline structure, whose patent belongs to UBE Chemical Industries [22,57], consists of a SiO_2 fiber mesh, which supports the TiO_2 catalyst, generating a TiO_2/SiO_2 catalyst/support system, avoiding the phenomenon of dragging of the photocatalyst from the surface of the support (a phenomenon known as peeling), as a consequence of its friction with the fluid of the liquid. The maximum pressure that the fiber can withstand is up to 10 kg/cm^2, with optimum performance in the range of 3–6 kg/cm^2. A TiO_2 semiconductor supported on a SiO_2 fiber, contained in 4 stainless steel conical meshes placed longitudinally on rods to immobilize them. The photocatalyst is located between the radiation source and the walls of the photoreactor [58]. Two low-pressure mercury lamps were used as irradiation sources: the first one, a 40 W lamp (Philips TUV 36T5 HE 4P SE UNP/32), emitting 15 W of UV-C ultraviolet radiation and the second one, a 75 W lamp (Philips TUV 36T5 HO 4P SE UNP/32), emitting 25 W of UV-C ultraviolet radiation, both with a maximum wavelength of 253.7 nm. Each lamp was placed inside a transparent quartz tube to prevent it from coming into contact with the sample.

3.2. Experimental Conditions

In order to establish the experimental conditions for the study of NH_4^+/NH_3 degradation in the UV-C photocatalytic reactor, the following considerations were taken into account.

Synthetic wastewater with NH_4^+/NH_3 concentrations similar to the outflows of leachate treatment from wastewater treatment plants was used [59] because its removal capacity is limited with NH_3-N > 100 mg/L wastewaters [60]. In addition, in order to study the different nitrogen species involved in the NH_4^+/NH_3 degradation mechanisms, possible inhibitions that may be produced by the adsorption of other species on the photocatalyst surface should be avoided [27,33].

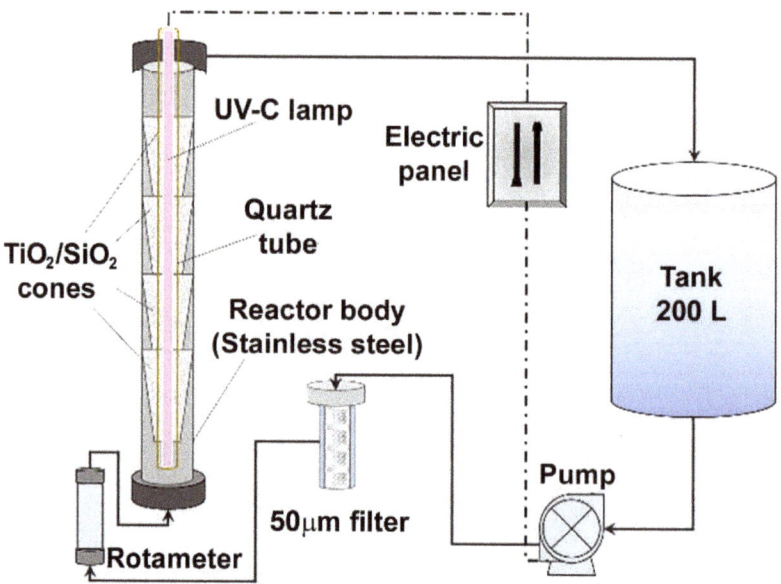

Figure 15. Photoreactor system used for photolysis and photocatalytic organic compounds degradation processes.

The experiments with synthetic wastewaters were carried out at pH 7.0, 9.0 and 11.0, proving that pH was an essential factor in the photolytic and photocatalytic degradation of NH_4^+/NH_3.

Another important operational parameter is the power of the UV-C ultraviolet irradiation and, therefore, the degradation experiments were carried out with two different lamps (15 W, 25 W), since the rate of photocatalytic activation and the formation of the electron-hole pair are strongly influenced by the power of UV-C irradiant light [34].

All experiments were carried out with 50 L of wastewater synthetic samples with initial ammonium concentrations (ammonium chloride provided by Sigma Aldrich) in the range 180–200 mg/L in the feed tank of the UV photoreactor. In the photocatalysis experiments, 4 cones of TiO_2/SiO_2 photocatalyst were used. For the photolysis experiments, the cones were taken out of the photoreactor. The pump is then switched on, adjusting the flow rate to 1000 L/h, maintaining a constant temperature at 20 ± 1 °C, by means of a refrigerant cooling coil, adjusting and maintaining constant pH (7.0, 9.0 and 11.0) with an 8% (m/v) NaOH aqueous solution and carrying out the experiments for 7–8 h under UV-C light. For each sample, in addition to temperature and pH, conductivity, dissolved oxygen, ammonium/ammonia, nitrite and nitrate concentration were measured by selective electrodes of the YSI6920 multiparameter probe. In the ammonium measurements throughout this work, the sum of the 2 conjugated NH_4^+/NH_3 species are expressed. Similarly, nitrogen species were determined by electronic absorption spectroscopy using a PG T80+ spectrophotometer and an HACH DR/2010 photometer. The stability studies of the photocatalytic fiber were carried out by energy dispersive X-ray elemental microanalysis (EDXMA), a technique which was also used to measure the titanium/silicon ratio of the fiber as well as the percentage of titanium on the surface fiber.

3.3. Data Analysis

Free open-source software was used for the different studies. For the chemical kinetics studies of the NH_4^+/NH_3 degradation tests, modelling and validation of the proposed mechanism, the statistical package SIMFIT was used. This package was developed at the University of Manchester by William G. Bardsley (http://www.simfit.org.uk) (accessed on 14 December 2021). The Spanish version of SIMFIT is maintained by F. J. Burguillo of the University of Salamanca (http://simfit.usal.es) (accessed on 16 December 2021) [61].

The "INRATE" program was used for the calculation of the initial velocities at the different substrate concentrations. The Langmuir–Hinshelwood integrated rate equation was fitted to the appropriate model using the "QNFIT" program and the differential equations were simultaneously fitted to the experimental data using the DQSOL program of the SIMFIT statistical package.

4. Conclusions

Under the conditions used, NH_4^+/NH_3 can be decomposed both by photolysis (UV radiation action) and photocatalysis (UV radiation plus photocatalytic fiber) routes, without neglecting the volatilization of ammonia by stripping processes (agitation and volatilization). It was found that the NH_4^+/NH_3 removal percentage is mainly a function of pH and lamp irradiation power, the higher the NH_3/NH_{4+} removal performance the higher the lamp irradiation power and the more basic the pH. An analysis of the nitrogen species occurring during the NH_4^+/NH_3 degradation process revealed that it decomposes mainly into nitrogen gas and nitrate, the intermediate species nitrite being very unstable, as it is rapidly formed and transformed into other species. Dissolved oxygen does not have a great influence on the reaction and remains almost constant. On the other hand, the conductivity increases with pH and lamp irradiation power, indicating that there is a greater formation of ionic intermediate species. The experimental data fit well to a Langmuir–Hinshelwood adsorption model. The low adsorption equilibrium constants (K = 0.0069) and the surface coverage values that for high NH_4^+/NH_3 concentrations do not exceed 10% coverage, show a low affinity of ammonium/ammonia for adsorption and surface reaction on the photocatalytic fiber, which translates into low operational rate constants (k_{op} = 0.13 h^{-1}). The photolysis rate constant being k_{PL} = 0.07 ± 0.01 h^{-1}.

The kinetic progress curves for NH_4^+/NH_3 degradation reaction tend towards plateau-type asymptotic values and not towards total NH_4^+/NH_3 removal, thus it could be thought that the reaction is inhibited by the adsorption of intermediate species on the surface of the photocatalytic fiber. By studying the initial reaction rates at different initial ammonia concentrations, it was shown that the decomposition of ammonia should tend to zero, or total removal, but the adsorption of intermediate species on the photocatalytic fiber, especially at low reaction rates (less basic pHs and lower lamp irradiation power) causes saturation of the fiber and explains the maximum limit of degradation. This fact was corroborated by the analytical semi-quantitative study (EDXMA) and the structural characterization by scanning electron microscopy of the photocatalytic fiber coating, which seems to confirm the hypothesis of adsorption of intermediate species on the photocatalyst surface.

The good fits of the experimental C(t) data to a model of two parallel NH_4^+/NH_3 decomposition reactions confirm the proposed NH_4^+/NH_3 degradation mechanism, which consists, on the one hand, in the formation of nitrogen gas and, on the other hand, in the formation of nitrate. At the optimal conditions, the rate constants for the disappearance of ammonia were k_3 = 0.154 h^{-1} and for the appearance of nitrate and nitrogen gas k_1 = 3.3 ± 0.2 10^{-5} h^{-1} and k_2 = 1.54 ± 0.07 10^{-1} h^{-1}, respectively.

This UV-C photocatalytic process is shown to be more effective for degrading NH_4^+/NH_3 than the photolytic process and does not require the addition of reagents, such as H_2O_2, for the formation of hydroxyl radicals and could compete with the processes carried out by adapted anammox bacteria in an SBR bioreactor. This comparative study will be the next target of future research.

Author Contributions: Methodology, J.C.G.-P. and M.G.-R.; validation, J.C.G.-P. and M.G.-R.; formal analysis, J.C.G.-P. and M.G.-R.; investigation, J.C.G.-P. and M.G.-R.; resources, J.C.G.-P., M.G.-R. and J.B.P.-N.; data curation, L.A.G.-B., M.G.-R. and J.C.G.-P.; writing—original draft preparation, J.C.G.-P. and M.G.-R.; writing—review and editing, J.C.G.-P., M.G.-R. and J.B.P.-N.; visualization, J.C.G.-P. and M.G.-R.; supervision, J.C.G.-P., L.A.G.-B., M.G.-R. and J.B.P.-N.; project administration, M.G.-R.; funding acquisition, J.C.G.-P. and M.G.-R. All authors have read and agreed to the published version of the manuscript.

Funding: This research was funded by the Consejo Nacional de Ciencia y Tecnología (CONACyT), Instituto Politécnico Nacional (IPN/SIP project 20190247and 20200670). This research also received the financial support and the supply of the photocatalytic fibre pilot plant from UBE Corporation Europe S.A. The content does not necessarily reflect the views and policies of the funding organizations.

Institutional Review Board Statement: Not applicable.

Informed Consent Statement: Not applicable.

Data Availability Statement: Data are contained within the article.

Acknowledgments: The authors wish to thank UBE Corporation Europe S.A. for supplying the photocatalytic fiber pilot plant and financially supporting this project. L.A.G.-B. and J.B.P.-N. thank the Consejo Nacional de Ciencia y Tecnología (CONACyT), who provided funding.

Conflicts of Interest: The authors declare no conflict of interest.

References

1. European Environment Agency (EEA). *NEC Directive Reporting Status*; EEA: Copenhagen, Denmark, 2019. [CrossRef]
2. Low, G.K.C.; McEvoy, S.R.; Matthews, R.W. Formation of nitrate and ammonium ions in titanium dioxide mediated photocatalytic degradation of organic compounds containing nitrogen atoms. *Environ. Sci. Technol.* **1991**, *25*, 460–467. [CrossRef]
3. Manahan, S.E. *Environmental Chemistry*, 1st ed.; Lewis Publishers: Chelsea, UK, 1991; pp. 128–132.
4. Randall, D.; Tsui, T. Ammonia toxicity in fish. *Mar. Pollut. Bull.* **2002**, *45*, 17–23. [CrossRef]
5. World Health Organization. *Guidelines for Drinking-Water Quality*, 2nd ed.; WHO: Geneva, Switzerland, 1996; Volume 2.
6. Agarwal, D.C. Stress corrosion in copper–nickel alloys: Influence of ammonia. *Br. Corros. J.* **2002**, *37*, 267–275. [CrossRef]
7. Huang, H.; Song, Q.; Wang, W.; Wu, S.; Dai, J. Treatment of anaerobic digester effluents of nylon wastewater through chemical precipitation and a sequencing batch reactor process. *J. Environ. Manag.* **2012**, *101*, 68–74. [CrossRef] [PubMed]
8. Huo, H.; Lin, H.; Dong, Y.; Cheng, H.; Wang, H.; Cao, L. Ammonia-nitrogen and phosphates sorption from simulated reclaimed waters by modified clinoptilolite. *J. Hazard. Mater.* **2012**, *229-230*, 292–297. [CrossRef]
9. Değermenci, N.; Ata, O.N.; Yildiz, E. Ammonia removal by air stripping in a semi-batch jet loop reactor. *J. Ind. Eng. Chem.* **2012**, *18*, 399–404. [CrossRef]
10. Stoquart, C.; Servais, P.; Bérubé, P.R.; Barbeau, B. Hybrid Membrane Processes using activated carbon treatment for drinking water: A review. *J. Membr. Sci.* **2012**, *411-412*, 1–12. [CrossRef]
11. Chauhan, R.; Srivastava, V.C. Electrochemical denitrification of highly contaminated actual nitrate wastewater by Ti/RuO$_2$ anode and iron cathode. *Chem. Eng. J.* **2019**, *386*, 122065. [CrossRef]
12. Mulder, A.; Van De Graaf, A.; Robertson, L.; Kuenen, J. Anaerobic ammonium oxidation discovered in a denitrifying fluidized bed reactor. *FEMS Microbiol. Ecol.* **1995**, *16*, 177–183. [CrossRef]
13. Wang, W.; Ding, Y.; Wang, Y.; Song, X.; Ambrose, R.F.; Ullman, J.L.; Winfrey, B.K.; Wang, J.; Gong, J. Treatment of rich ammonia nitrogen wastewater with polyvinyl alcohol immobilized nitrifier biofortified constructed wetlands. *Ecol. Eng.* **2016**, *94*, 7–11. [CrossRef]
14. Zhang, G.; Ruan, J.; Du, T. Recent Advances on Photocatalytic and Electrochemical Oxidation for Ammonia Treatment from Water/Wastewater. *ACS EST Eng.* **2020**, *1*, 310–325. [CrossRef]
15. Ahn, Y.-H. Sustainable nitrogen elimination biotechnologies: A review. *Process Biochem.* **2006**, *41*, 1709–1721. [CrossRef]
16. Shavisi, Y.; Sharifnia, S.; Hosseini, S.; Khadivi, M. Application of TiO$_2$/perlite photocatalysis for degradation of ammonia in wastewater. *J. Ind. Eng. Chem.* **2013**, *20*, 278–283. [CrossRef]
17. Bahadori, E.; Conte, F.; Tripodi, A.; Ramis, G.; Rossetti, I. Photocatalytic Selective Oxidation of Ammonia in a Semi-Batch Reactor: Unravelling the Effect of Reaction Conditions and Metal Co-Catalysts. *Catalysts* **2021**, *11*, 209. [CrossRef]
18. Zendehzaban, M.; Sharifnia, S.; Hosseini, S.N. Photocatalytic degradation of ammonia by light expanded clay aggregate (LECA)-coating of TiO2 nanoparticles. *Korean J. Chem. Eng.* **2013**, *30*, 574–579. [CrossRef]
19. Feng, J.; Zhang, X.; Zhang, G.; Li, J.; Song, W.; Xu, Z. Improved photocatalytic conversion of high−concentration ammonia in water by low−cost Cu/TiO2 and its mechanism study. *Chemosphere* **2021**, *274*, 129689. [CrossRef] [PubMed]
20. Mozzanega, H.; Herrmann, J.M.; Pichat, P. Ammonia oxidation over UV-irradiated titanium dioxide at room temperature. *J. Phys. Chem.* **1979**, *83*, 2251–2255. [CrossRef]

21. Dijkstra, M.; Michorius, A.; Buwalda, H.; Panneman, H.; Winkelman, J.; Beenackers, A. Comparison of the efficiency of immobilized and suspended systems in photocatalytic degradation. *Catal. Today* **2001**, *66*, 487–494. [CrossRef]
22. Ishikawa, T.; Yamaoka, H.; Harada, Y.; Fujii, T.; Nagasawa, T. A general process for in situ formation of functional surface layers on ceramics. *Nature* **2002**, *416*, 64–67. [CrossRef] [PubMed]
23. Li, Y.-N.; Chen, Z.-Y.; Bao, S.-J.; Wang, M.-Q.; Song, C.-L.; Pu, S.; Long, D. Ultrafine TiO_2 encapsulated in nitrogen-doped porous carbon framework for photocatalytic degradation of ammonia gas. *Chem. Eng. J.* **2018**, *331*, 383–388. [CrossRef]
24. Zhou, Q.; Yin, H.; Wang, A.; Si, Y. Preparation of hollow B–SiO_2@TiO_2 composites and their photocatalytic performances for degradation of ammonia-nitrogen and green algae in aqueous solution. *Chin. J. Chem. Eng.* **2019**, *27*, 2535–2543. [CrossRef]
25. Peng, X.; Wang, M.; Hu, F.; Qiu, F.; Dai, H.; Cao, Z. Facile fabrication of hollow biochar carbon-doped TiO_2/CuO composites for the photocatalytic degradation of ammonia nitrogen from aqueous solution. *J. Alloy. Compd.* **2018**, *770*, 1055–1063. [CrossRef]
26. Wang, J.; Liang, X.; Chen, P.; Zhang, D.; Yang, S.; Liu, Z. Microstructure and photocatalytic properties of Ag/Ce^{4+}/La^{3+} co-modified TiO_2/Basalt fiber for ammonia–nitrogen removal from synthetic wastewater. *J. Sol-Gel Sci. Technol.* **2016**, *82*, 289–298. [CrossRef]
27. Bravo, A.; Garcia, J.; Domenech, X.; Peral, J. Some Aspects of the Photocatalytic Oxidation of ammonium ion by Titanium Dioxide. *J. Chem. Res.* **1993**, *1*, 376–377.
28. Bonsen, E.-M.; Schroeter, S.; Jacobs, H.; Broekaert, J. Photocatalytic degradation of ammonia with t102 as photocatalyst in the laboratory and under the use of solar radiation. *Chemosphere* **1997**, *35*, 1431–1445. [CrossRef]
29. Zhu, X.; Castleberry, S.R.; Nanny, M.A.; Butler, E.C. Effects of pH and Catalyst Concentration on Photocatalytic Oxidation of Aqueous Ammonia and Nitrite in Titanium Dioxide Suspensions. *Environ. Sci. Technol.* **2005**, *39*, 3784–3791. [CrossRef]
30. Sun, D.; Sun, W.; Yang, W.; Li, Q.; Shang, J.K. Efficient photocatalytic removal of aqueous NH_4^+–NH_3 by palladium-modified nitrogen-doped titanium oxide nanoparticles under visible light illumination, even in weak alkaline solutions. *Chem. Eng. J.* **2015**, *264*, 728–734. [CrossRef]
31. Altomare, M.; Dozzi, M.V.; Chiarello, G.L.; Di Paola, A.; Palmisano, L.; Selli, E. High activity of brookite TiO_2 nanoparticles in the photocatalytic abatement of ammonia in water. *Catal. Today* **2015**, *252*, 184–189. [CrossRef]
32. Shibuya, S.; Sekine, Y.; Mikami, I. Influence of pH and pH adjustment conditions on photocatalytic oxidation of aqueous ammonia under airflow over Pt-loaded TiO_2. *Appl. Catal. A Gen.* **2015**, *496*, 73–78. [CrossRef]
33. Vohra, M.; Selimuzzaman, S.; Al-Suwaiyan, M. NH_4^+-NH_3 removal from simulated wastewater using UV-TiO_2 photocatalysis: Effect of co-pollutants and pH. *Environ. Technol.* **2010**, *31*, 641–654. [CrossRef] [PubMed]
34. Cassano, A.E.; Alfano, O.M. Reaction engineering of suspended solid heterogeneous photocatalytic reactors. *Catal. Today* **2000**, *58*, 167–197. [CrossRef]
35. Lee, K.M.; Lai, C.W.; Ngai, K.S.; Juan, J.C. Recent developments of zinc oxide based photocatalyst in water treatment technology: A review. *Water Res.* **2016**, *88*, 428–448. [CrossRef] [PubMed]
36. Altomare, M.; Chiarello, G.L.; Costa, A.; Guarino, M.; Selli, E. Photocatalytic abatement of ammonia in nitrogen-containing effluents. *Chem. Eng. J.* **2012**, *191*, 394–401. [CrossRef]
37. Serewicz, A.; Noyes Aibert, W., Jr. The Photolysis of Ammonia in the Presence of Nitric Oxide. *J. Phys. Chem.* **1959**, *63*, 843–845. [CrossRef]
38. Huang, L.; Li, L.; Dong, W.; Liu, Y.; Hou, H. Removal of Ammonia by OH Radical in Aqueous Phase. *Environ. Sci. Technol.* **2008**, *42*, 8070–8075. [CrossRef]
39. Wang, J.; Song, M.; Chen, B.; Wang, L.; Zhu, R. Effects of pH and H_2O_2 on ammonia, nitrite and nitrate transformations during UV254nm irradiation: Implications to nitrogen removal and analysis. *Chemosphere* **2017**, *184*, 1003–1011. [CrossRef]
40. Wang, S.; Ye, Z.; Taghipour, F. UV photoelectrochemical process for the synergistic degradation of total ammonia nitrogen (TAN). *J. Clean. Prod.* **2020**, *289*, 125645. [CrossRef]
41. Gerischer, H.; Mauerer, A. Untersuchungen Zur anodischen Oxidation von Ammoniak an Platin-Elektroden. *J. Electroanal. Chem. Interfacial Electrochem.* **1970**, *25*, 421–433. [CrossRef]
42. Murgia, S.M.; Poletti, A.; Selvaggi, R. Photocatalytic Degradation of High Ammonia Concentration Water Solutions by TiO_2. *Ann. Chim.* **2005**, *95*, 335–343. [CrossRef]
43. Lee, J.; Park, H.; Choi, W. Selective Photocatalytic Oxidation of NH_3 to N_2 on Platinized TiO_2 in Water. *Environ. Sci. Technol.* **2002**, *36*, 5462–5468. [CrossRef]
44. Núñez-Núñez, C.M.; Chairez-Hernández, I.; García-Roig, M.; García-Prieto, J.C.; Melgoza-Alemán, R.M.; Proal-Nájera, J.B. UV-C/H_2O_2 heterogeneous photocatalytic inactivation of coliforms in municipal wastewater in a TiO_2/SiO_2 fixed bed reactor: A kinetic and statistical approach. *React. Kinet. Mech. Catal.* **2018**, *125*, 1159–1177. [CrossRef]
45. Parks, G.A. The Isoelectric Points of Solid Oxides, Solid Hydroxides, and Aqueous Hydroxo Complex Systems. *Chem. Rev.* **1965**, *65*, 177–198. [CrossRef]
46. Biyoghe Bi Ndong, L.; Ibondou, M.P.; Gu, X.; Lu, S.; Qiu, Z.; Sui, Q.; Mbadinga, S.M. Enhanced Photocatalytic Activity of TiO_2 Nanosheets by Doping with Cu for Chlorinated Solvent Pollutants Degradation. *Ind. Eng. Chem. Res.* **2014**, *53*, 1368–1376. [CrossRef]
47. Gong, X.; Wang, H.; Yang, C.; Li, Q.; Chen, X.; Hu, J. Photocatalytic degradation of high ammonia concentration wastewater by TiO_2. *Futur. Cities Environ.* **2015**, *1*, 12. [CrossRef]

48. Herrmann, J.M.; Mu, W.; Pichat, P. *Heterogeneous catalysis and fine chemicals II*; Guisnet, M., Ed.; Elsevier: Amsterdam, The Netherlands, 1991; pp. 405–420.
49. Szabó-Bárdos, E.; Czili, H.; Horváth, A. Photocatalytic oxidation of oxalic acid enhanced by silver deposition on a TiO_2 surface. *J. Photochem. Photobiol. A Chem.* **2003**, *154*, 195–201. [CrossRef]
50. Zhu, X.; Nanny, M.A.; Butler, E.C. Effect of inorganic anions on the titanium dioxide-based photocatalytic oxidation of aqueous ammonia and nitrite. *J. Photochem. Photobiol. A Chem.* **2007**, *185*, 289–294. [CrossRef]
51. Mack, J.; Bolton, J.R. Photochemistry of nitrite and nitrate in aqueous solution: A review. *J. Photochem. Photobiol. A Chem.* **1999**, *128*, 1–13. [CrossRef]
52. Doudrick, K.; Yang, T.; Hristovski, K.; Westerhoff, P. Photocatalytic nitrate reduction in water: Managing the hole scavenger and reaction by-product selectivity. *Appl. Catal. B Environ.* **2013**, *136–137*, 40–47. [CrossRef]
53. Krasae, N.; Wantala, K. Enhanced nitrogen selectivity for nitrate reduction on Cu–nZVI by TiO_2 photocatalysts under UV irradiation. *Appl. Surf. Sci.* **2016**, *380*, 309–317. [CrossRef]
54. García-Prieto, J.C.; González-Burciaga, L.A.; Proal-Nájera, J.B.; García-Roig, M. Study of Influence Factors in the Evaluation of the Performance of a Photocatalytic Fibre Reactor (TiO_2/SiO_2) for the Removal of Organic Pollutants from Water. *Catalysts* **2022**, *12*, 122. [CrossRef]
55. Cho, S.; Kambey, C.; Nguyen, V.K. Performance of Anammox Processes for Wastewater Treatment: A Critical Review on Effects of Operational Conditions and Environmental Stresses. *Water* **2019**, *12*, 20. [CrossRef]
56. Cunningham, J.; Al-Sayyed, G.; Srijaranai, S. *Adsorption of Model Pollutants on TiO_2 Particles in Relation to Photoremediation of Contaminated Water*; Helz, G., Zepp, R., Crosby, D., Eds.; Aquatic and Surface Chemistry: Boca Raton, FL, USA, 1994; pp. 317–348.
57. Ishikawa, T. Advances in inorganic fibres. In *Polymeric and Inorganic Fibres*; Springer: Berlin, Heidelberg, 2005; Volume 178, pp. 109–144. [CrossRef]
58. Cachaza Silverio, E.; Diez Mateos, A.; García-Prieto, J.C. La fotocatálisis como tecnología para la desinfección del agua sin aditivos químicos. Ensayos del reactor fotocatalítico UBE en la inactivación de microorganismos. *Tecnología del Agua*. **2005**, *261*, 81–85.
59. Mehta, C.M.; Khunjar, W.O.; Nguyen, V.; Tait, S.; Batstone, D. Technologies to Recover Nutrients from Waste Streams: A Critical Review. *Crit. Rev. Environ. Sci. Technol.* **2014**, *45*, 385–427. [CrossRef]
60. Li, Y.; Guo, A.J.; Zhou, J. Research of Treating High Ammonia-N Wastewater by CASS Process. *Adv. Mater. Res.* **2011**, *250–253*, 3844–3847. [CrossRef]
61. Bardsley, W.G. Simfit Statistical package. v. 7.3.1 Academic 64-bit University of Manchester UK. Available online: http://www.simfit.man.ac.uk (accessed on 14 December 2021).

Article

Block Copolymer and Cellulose Templated Mesoporous TiO$_2$-SiO$_2$ Nanocomposite as Superior Photocatalyst

Sudipto Pal [1,*], Antonietta Taurino [2], Massimo Catalano [2] and Antonio Licciulli [1,3]

[1] Department of Engineering for Innovation, University of Salento, Via per Monteroni, 73100 Lecce, Italy; antonio.licciulli@unisalento.it
[2] Institute for Microelectronics and Microsystems (IMM), CNR, Via Monteroni, 73100 Lecce, Italy; antonietta.taurino@le.imm.cnr.it (A.T.); massimo.catalano@le.imm.cnr.it (M.C.)
[3] Institute of Nanotechnology, CNR Nanotec, Consiglio Nazionale Delle Ricerche, Via Monteroni, 73100 Lecce, Italy
* Correspondence: sudipto.pal@unisalento.it

Abstract: A dual soft-templating method was developed to produce highly crystalline and mesoporous TiO$_2$-SiO$_2$ nanocomposites. Pluronic F127 as the structure-directing agent and pure cellulose as the surface area modifier were used as the templating media. While Pluronic F127 served as the sacrificing media for generating a mesoporous structure in an acidic pH, cellulose templating helped to increase the specific surface area without affecting the mesoporosity of the TiO$_2$-SiO$_2$ nanostructures. Calcination at elevated temperature removed all the organics and formed pure inorganic TiO$_2$-SiO$_2$ composites as revealed by TGA and FTIR analyses. An optimum amount of SiO$_2$ insertion in the TiO$_2$ matrix increased the thermal stability of the crystalline anatase phase. BET surface area measurement along with low angle XRD revealed the formation of a mesoporous structure in the composites. The photocatalytic activity was evaluated by the degradation of Rhodamine B, Methylene Blue, and 4-Nitrophenol as the model pollutants under solar light irradiation, where the superior photo-degradation activity of Pluronic F127/cellulose templated TiO$_2$-SiO$_2$ was observed compared to pure Pluronic templated composite and commercial Evonik P25 TiO$_2$. The higher photocatalytic activity was achieved due to the higher thermal stability of the nanocrystalline anatase phase, the mesoporosity, and the higher specific surface area.

Keywords: TiO$_2$-SiO$_2$; mesoporous; F127; high surface area; solar photocatalysis; dye degradation

Citation: Pal, S.; Taurino, A.; Catalano, M.; Licciulli, A. Block Copolymer and Cellulose Templated Mesoporous TiO$_2$-SiO$_2$ Nanocomposite as Superior Photocatalyst. *Catalysts* **2022**, *12*, 770. https://doi.org/10.3390/catal12070770

Academic Editors: Magdalena Zybert and Katarzyna Antoniak-Jurak

Received: 16 May 2022
Accepted: 8 July 2022
Published: 12 July 2022

Publisher's Note: MDPI stays neutral with regard to jurisdictional claims in published maps and institutional affiliations.

Copyright: © 2022 by the authors. Licensee MDPI, Basel, Switzerland. This article is an open access article distributed under the terms and conditions of the Creative Commons Attribution (CC BY) license (https:// creativecommons.org/licenses/by/ 4.0/).

1. Introduction

Water contamination from the wastewater released by the textile industries, pharmaceutical plants, and agrochemical and leather processing factories is one of the biggest concerns among the various forms of environmental pollution [1,2]. A large number of organic dyes are currently used for various purposes, e.g., Rhodamine B and Methylene Blue as colorants and 4-Nitrophenol for manufacturing drugs and pesticides. Most of them are highly water soluble and non-biodegradable, and therefore difficult to separate by filtering processes. Among the various degradation techniques, photocatalytic mineralization is one of the best environmental-friendly techniques, since other chemical/biological processes can produce secondary byproducts [3]. Titanium dioxide (TiO$_2$) with a stable anatase crystalline phase is considered to be one of the best photocatalyst materials, which is widely used in the photodecomposition of organic pollutants, wastewater treatment, and environmental remediations [1,4–7]. A higher photocatalytic activity, long term photo and chemical stability, low toxicity, and relatively lower cost make it an excellent candidate in heterogeneous photocatalysis [8,9]. The basic principle of photocatalysis in TiO$_2$ relies on the formation of excitons (electron-hole pairs) generated by the excitation of the absorbed photon energy that is greater than the band gap energy of TiO$_2$ (3.0–3.2 eV) and their migration to the catalyst surface. These photogenerated excitons may take place in the redox reactions where superoxide radical anions ($O_2^{\bullet -}$) and hydroxyl radicals ($OH^{\bullet}$) are produced in the presence of oxygen and water, and later on in the process end up with the mineralization of the organic species adsorbed on TiO_2 surfaces [10–12]. The photocatalytic efficiency of TiO_2 depends on several factors such as crystalline phase, particle size, specific surface area, and porosity [13]. The high crystallinity in TiO_2 enhances the generation and migration of the photogenerated excitons, whereas its higher surface area and mesoporosity help to enhance its reactivity by means of increasing the active sites [10,14]. Moreover, the higher surface area along with mesoscale porosity could trap the contaminants/molecules into their porous and well-connected network, which could initiate the photocatalytic reaction process very quickly.

The soft-templating route wherein non-ionic (such as triblock copolymers, Pluronic F127, P123) [15] or ionic (such as CTAB) [16] surfactants are used as the scarifying media for regular pore generation is the most popular way to synthesize mesoporous metal oxide nanomaterials. After burning out these surfactants at a certain temperature, the mesoporous structure is retained by the synthesized oxide nanomaterial. The mesoporous nanocrystalline TiO_2, either in powder form or thin film, synthesized following this route has been reported [17–19]. However, due to the prolonged calcination process at elevated temperatures, which are necessary to obtain the mesoporous structure, TiO_2 often suffers from the anatase to rutile phase transformation (due to thermal instability of the anatase phase) that hinders its photocatalytic activity [18,20–22]. This could be avoided by incorporating silica (SiO_2) into the titania matrices [14,23–25]. Silica insertion into the TiO_2 nanostructure not only increases the thermal stability of the highly photoactive anatase phase but also helps to prevent the mesoporous structure from collapsing [2,26–28]. Introducing cellulose matrix as the second templating media to the triblock copolymer-TiO_2-SiO_2 composite could further enhance the specific surface area and support the mesoporous structure with a higher degree of dispersibility in aqueous media that would enhance the photocatalytic efficiency [29]. Due to the 3D web-like nanofibrous structure of cellulose, there has been significant interest in synthesizing cellulose-TiO_2 nanocomposites, either by the immobilization of TiO_2 on cellulose or by using as template [29–33]. However, most of the preparation methods need either multistep stages or take a prolonged period to obtain the composite mesoporous nanostructure. In this work, we have demonstrated a quick procedure to anchor the TiO_2-SiO_2-triblock copolymer composite on a cellulose (commercial filter paper) matrix and obtain pure inorganic TiO_2-SiO_2 mesoporous nanostructure after burning out the cellulose template and the copolymer as well. In this unique templating method, the triblock copolymer acts as the structure-directing and mesopore-generating agent, whereas the SiO_2 counterpart fixes the thermal stability of the anatase phase and the mesostructure, and finally the cellulose templating enhances the specific surface area and porosity. The obtained TiO_2-SiO_2 nanocomposites showed excellent thermal stability of the anatase phase and a higher photocatalytic efficiency as compared to the commercial P25 TiO_2.

2. Materials and Methods
2.1. Preparation of the Photocatalyst

All the reagents involved in the catalyst synthesis were used as received without making any further modifications. Triblock copolymer Pluronic F127 (EO_{106} PO_{70} EO_{106}, average molecular weight 12.6 kDa, Sigma-Aldrich, Saint Louis, MO, USA), titanium tetraisopropoxide ($Ti(O^iPr)_4$, TTIP, 97%), tetraethoxysilane ($Si(OC_2H_5)_4$, TEOS, 97%, Sigma-Aldrich, Saint Louis, MO, USA), hydrochloric acid (HCl, 37–38%, J.T. Baker), and Whatman filter paper (qualitative, grade 595) were used as the reagents. Commercial P25 TiO_2 nanopowder was purchased from Evonik Resource Efficiency GmbH (Hanau-Wolfgang, Hesse, Germany). First, TiO_2-SiO_2 composites with different TiO_2/SiO_2 weight ratios were synthesized according to our previous work with a little modification [34]. As in a typical synthesis of TS82, the required amount of F127 (0.012 M of oxides) was dissolved in 600 g of 2 M HCl and 150 g of water with vigorous stirring. After obtaining a clear solution, TEOS was added drop wise followed by stirring for 1 h and then TTIP was slowly added to the mixture followed by overnight stirring for a hydrolysis-condensation reaction. Then, the mixed sol was transferred to a polypropylene bottle and kept in an oven at 80 °C for 48 h. Then, the solid precipitate was separated and dried to form the xerogel, followed by calcination at 550 °C for 6 h with a heating and cooling rate of 1 °C/min to remove the organic contents. It is to be noted that the highest photocatalytic efficiency was achieved with the TS82 sample [35] (Figure S3, Table S2), so only this composition was chosen to prepare the cellulose-TiO_2/SiO_2 nanocomposite, which is denoted as TS82C throughout this article. To obtain this, the filter papers were cut into pieces and impregnated during the hydrolysis–condensation process and followed the procedure as described above with the similar heating cycle. After removal of all organics, thin flakes of TS82C with bright white colour were obtained.

2.2. Characterizations

Crystalline phases of the nanocomposite powder samples were characterized by wide angle (10–80° 2θ) X-ray diffraction (XRD) spectrometry performed on a Rigaku Ultima X-ray diffractometer using CuKα radiation (λ = 1.5406 Å) operating at 40 kV/30 mA with a step size of 0.02°. The low angle (0.3–10° 2θ) XRD pattern (GIXRD) was collected with a Rigaku SmartLab diffractometer operating at 9 kW. Thermogravimetric analysis (TGA) was carried out with a Mettler thermo-analyzer (Mettler Toledo, Star system) at a heating rate of 5 °C/min in an air atmosphere. FTIR spectra of the obtained composite was carried out with a JASCO FTIR-6300 over the range of 4000–400 cm^{-1} with a resolution of 4 cm^{-1} and accumulating 256 scans for each measurement adopting the KBr disc method. Raman spectral measurements (FT-Raman) of the powders were performed on a JASCO RFT-6000 Raman attachment by using a 1064 nm CW 500 mW laser source and spectral resolution of 4 cm^{-1}. Surface area and porosity of the samples were measured from nitrogen adsorption–desorption isotherms at liquid nitrogen temperature (77 K) by using a Quantachrome NOVA 2200e surface area and pore size analyzer. Before each measurement, the powder samples were degassed overnight at 423 K under nitrogen flow. The specific surface area was calculated by the multipoint BET (Brunauer–Emmett–Teller) equation from the N_2 adsorption branch of the isotherm in the relative pressure range of 0.05–0.35. The pore size distribution was calculated using the Barret–Joyner–Halenda (BJH) method from the desorption branch of the isotherm. FESEM measurements were performed with a Zeiss Sigma VP (Carl Zeiss, Jena, Germany) field emission scanning electron microscope. Transmission electron microscopic (TEM) analyses were performed with a JEOL JEM-1011 transmission electron microscope operating at 100 kV and equipped with a 7.1-megapixel CCD camera (Orius SC1000, Gatan, Pleasanton, CA, USA). The TEM micrographs were processed with Gatan's Digital Micrograph (DM) software. Average particle size was determined by counting 50 particles from two different micrographs.

2.3. Photocatalytic Experimental Set Up

Photocatalytic efficiency of the synthesized nanocomposites was tested by observing the degradation of rhodamine B (RhB), methylene blue (MB), and 4-Nitrophenol (4NP) aqueous solution under solar light irradiation (A single 300 W tungsten lamp with spectral irradiance of 13.6 W/m^2 and 41.4 W/m^2 in the wavelength range of 315–400 nm and 380–780 nm, respectively). The distance from the bottom of the lamp to the upper level of the dye solution was maintained at about 30 cm. In each case, prior to the solar light exposure, the composite powder samples (1 g/L) were dispersed in respective aqueous solutions of the dye molecules (200 mL, 15×10^{-6} M) and stirring was continued for 18 h in dark conditions to ensure the adsorption–desorption equilibrium. Then, they were kept under solar light irradiation and stirred constantly, and aliquot amounts (1 mL) of the irradiated solutions were extracted at 5 min intervals. Photocatalytic decomposition was monitored by measuring the absorption band of the respective dye solutions with an Agilent Cary 5000 series UV-Visible spectrophotometer. In all cases, prior to the optical measurement, the catalysts were separated from the solution by high-speed centrifugation.

3. Results and Discussion

3.1. Thermogravimetric Analyses and FTIR

The thermal decomposition behavior of the cellulose-titania/silica composite was investigated by thermogravimetric analysis (TGA), which is presented in Figure 1a. The TGA curve shows a gradual weight loss with an increasing calcination temperature. A small weight loss (3.45%) below 110 °C is seen at the first step, which is due to the loss of some volatile species such as water, ethanol/propanol, and HCl [19,36]. Then, an 18.8% weight loss was observed in the temperature range of 200–254 °C, which stems from the decomposition of the F127 template [20,37]. The steep decrease (weight loss 49.61%) up to 340 °C could be due to the carbonization of the cellulose template [30,32,38]. The weight loss (13.64%) at the final step from 340–420 °C could be attributed to the decomposition of some residual hydroxyl group or oxidation of the carbonaceous species and possibly the transformation of amorphous titania to anatase phase [19,38]. After that, no changes in weight loss were observed, indicating the complete removal of both the templates and high temperature stability of the nanocomposite. As we have performed thermal treatment at 550 °C, no organic residue was present in our sample.

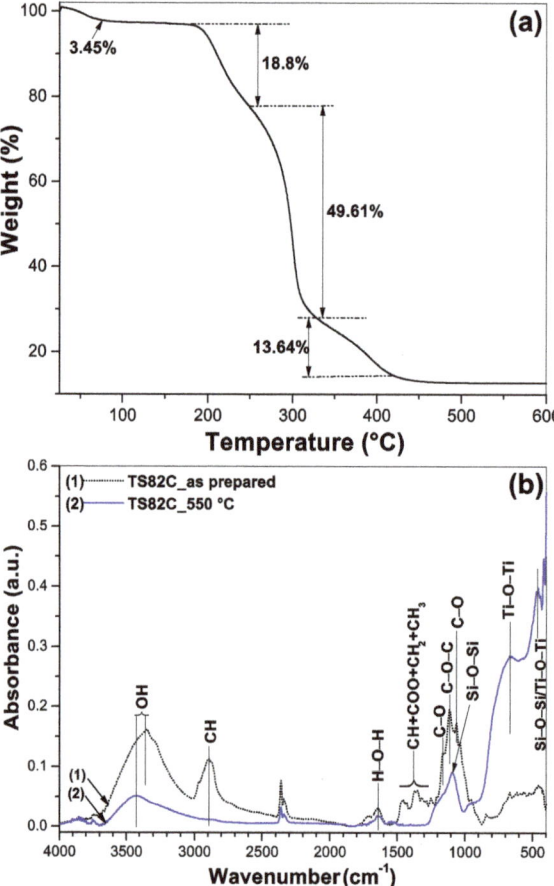

Figure 1. (a) Thermogravimetric analyses (TGA) of as prepared TS82C. (b) FTIR spectra of as prepared (curve 1) and 550 °C annealed (curve 2) TS82C.

The elimination of the organic templates from the composite sample after the calcination and determination of the structural information of TiO_2-SiO_2 was further supported by FTIR measurement. The FTIR spectra of the as prepared composite and after calcination at 550 °C are shown in Figure 1b. The as prepared sample (curve 1) shows a strong band around 3400 cm^{-1} due to the stretching vibration of hydroxyl groups coming from both the cellulose and TiO_2/SiO_2 matrices [30,37,39]. Other bands appeared at 2886, 1240–1490, 1158, and 1056 cm^{-1} could be assigned to CH, CH/COO/CH_2/CH_3, CO, and CO vibrations of the cellulose template [40,41]. It is noteworthy that the Si–O–Si asymmetric stretching band at 1086 cm^{-1} is superimposed by the strong C–O–C stretching vibration at 1110 cm^{-1} arising from the Pluronic template [42,43]. All the bands related to the organic species are indicated in Figure 1b. After annealing at 550 °C, all the bands related to the organic species disappeared, confirming their decomposition from the composite. At this stage, a new wide band centered at 668 cm^{-1} arose, which can be assigned to the characteristic stretching vibration of Ti–O–Ti coming from TiO_2 [39,44,45]. Another band at 465 cm^{-1} could be due to the Si–O–Si/Ti–O–Ti network of the composite. The strong appearance of the inorganic framework (Si/Ti–O–Si/Ti) at the lower wavenumber indicates the high degree of condensation in the inorganic network after the heat treatment [37].

3.2. Nanocrystalline Phase Composition and Mesoporosity

The crystalline nature and corresponding phase composition of the mixed nanocomposite was investigated using the powder XRD diffraction method and is shown in Figure 2a. All the diffraction peaks are assigned with their respective '*hkl*' parameters and were indexed as 25.32° (101), 36.98° (103), 37.84° (004), 38.56° (112), 48.05° (200), 53.98° (105), 55.12° (211), 62.79° (204), 68.84° (116), 70.31° (220), 75.15° (215), and 76.06° (301), which correspond to their pure anatase crystalline phases [45–47] (JCPDS No 84-1286). The average crystalline size calculated from XRD (Using Scherrer's formula, considering the strongest diffraction peak related to the 101 plane) was found to be 15.4 nm. The inset of Figure 2 shows the low-angle XRD pattern of the heat-treated sample where a strong reflection appears along with two other weaker reflections in the 2θ range of 0.5–2.5° with *d* spacings of 139.6, 98.07, and 68.96 Å. These peaks have *d* spacing ratios of $\sim\sqrt{2}:\sqrt{4}:\sqrt{8}$, which can be indexed as (110), (200), and (220) reflections, respectively, corresponding to the cubic $Im\bar{3}m$ space group [43,48,49] with the lattice constant *a* = 197.4 Å. These data confirm the formation of SBA-16 type cubic mesoporous structures in our samples, which are expected when Pluronic F127 ($EO_{106}PO_{70}EO_{106}$) triblock copolymer is used as a structure-directing agent.

The result obtained in the XRD regarding the acquisition of a pure anatase crystalline phase is further supported by the Raman spectral measurement of the TS82C powder sample heat treated at 550 °C, shown in Figure 2b. The spectrum shows strong and well-resolved bands at 150, 201, 401, 516, 520 (superimposed with 516 cm^{-1} band), and 643 cm^{-1}, which can be attributed to the six characteristic Raman-active modes of anatase crystalline phase with the symmetries of E_g, E_g, B_{1g}, A_{1g}/B_{1g}, and E_g, respectively [46,48,50]. No other bands were observed, due to either the rutile or brookite crystalline phases [45].

The mesoporosity of the calcined TiO_2/SiO_2 nanocomposites was investigated by Brunauer–Emmett–Teller (BET) surface area measurements. The N_2 adsorption–desorption isotherms along with BJH pore-size distribution of calcined TS82 and TS82C are presented in Figure 3. The data for the commercial P25 TiO_2 powder is also shown for comparison. The full sets of data corresponding to the BET characterization of the mixed oxides with varying TiO_2/SiO_2 ratios are provided as Supporting Information (Figure S1, Table S1). All the plots show a type IV N_2 sorption isotherm, which is characteristic of the mesoporous structure [44,51–53]. Interestingly, the pure silica sample (Figure S1, sample TS01) displayed a more prominent H1 type hysteresis with a steep slope at a higher relative pressure due to capillary condensation in the mesopores, reflecting the formation of highly ordered mesopores and interparticle voids between the primary particles [52,54]. However, when increasing the amount of TiO_2 the hysteresis loop became wider, which is believed to be due to the presence of heterogeneous mesopores. This is evidenced by the multimodal appearance and broadening of the BJH pore size distribution with the increasing amount of TiO$_2$ content (Figure S1, plot b). The pore size increased from 3.42 to 16.72 nm for the TS01 (SiO$_2$) and TS10 (TiO$_2$) samples, respectively. Consequently, the specific surface area also decreased from 553.31 (TS01) to 53.97 m^{2}g^{-1} (TS10) with the increasing wt% of TiO$_2$ (Table S1), which can be explained by the expansion of the pore size along with an increasing crystal size (Wide angle XRD analyses showed crystal size of 15.74 nm for TS73 and 17.55 nm for TS10) [52]. Now, if we compare BET analyses between TS82 and TS82C, the advantage of cellulose templating can be directly evaluated. TS82 displayed a specific surface area of 165.93 m^{2}g^{-1}, whereas for TS82C it was 186.82 m^{2}g^{-1}. Any significant changes in pore size distribution or pore volume were not noticed as reported in Table 1. So, the cellulose templating acted as the enhancer of the surface area of the composite, which in fact increased the photocatalytic efficiency (discussed later).

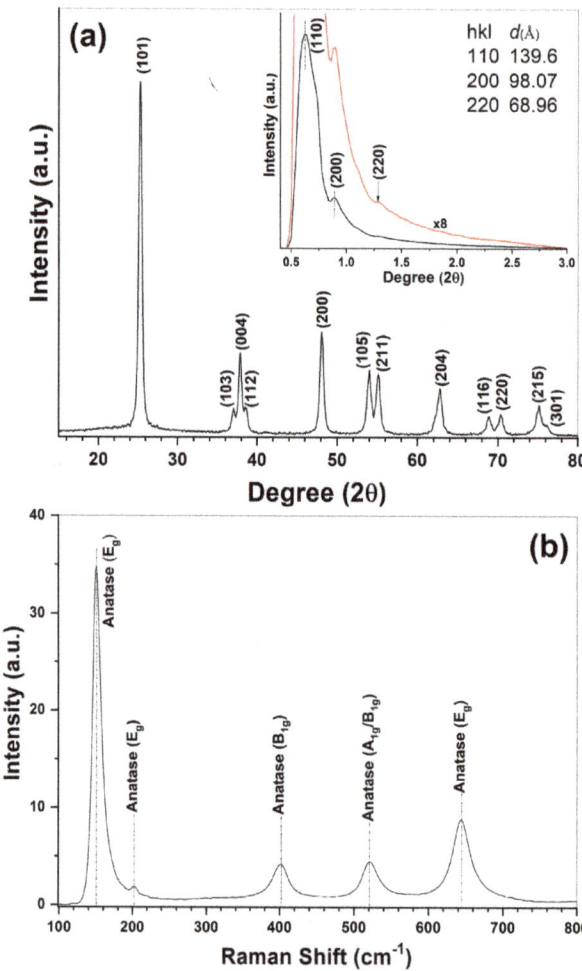

Figure 2. (a) Powder XRD pattern of the TS82C nanocomposite annealed at 550 °C. The inset shows low angle XRD pattern, where the hkl planes of the corresponding diffraction peaks along with the lattice spacings (*d*-values) are also shown. (b) Raman spectra of the TS82C powder calcined at 550 °C, showing different characteristic bands of the anatase crystalline phase.

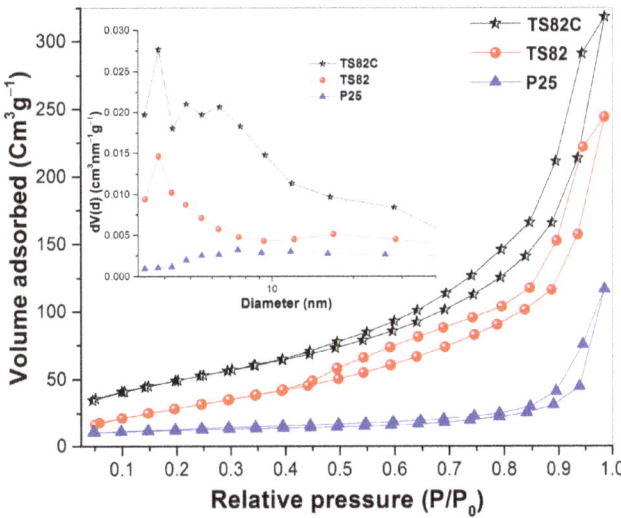

Figure 3. Nitrogen adsorption–desorption isotherm and BJH pore-size distribution plots (shown in inset) of TS82, TS82C nanocomposite calcined at 550 °C and P25 commercial (Evonik) TiO$_2$ powder.

Table 1. BET surface area parameters of the respective samples.

Sample	^{a} S$_{BET}$ (m^{2}g^{-1})	^{b} D$_{av}$ (nm)	^{c} V (cm^{3}g^{-1})	^{d} W$_{xrd}$		^{e} D$_{xrd}$
				W$_A$	W$_R$	
TS82C	186.82	3.84	0.41	1.0	0	15.40
TS82	165.93	3.78	0.38	1.0	0	15.99
P25	38.25	7.61	0.17	0.84	0.16	22.16

^{a} Specific surface area derived from the adsorption isotherm (P/P$_0$, 0.10–0.35). $^{b,\,c}$ Average pore diameter and pore volume calculated using BJH method from the desorption isotherm. ^{d} Anatase (W$_A$) and rutile (W$_R$) crystalline phase composition obtained from xrd analyses: W$_A$ = [1 + 1.26 (I$_{R110}$/I$_{A101}$)]$^{-1}$; W$_R$ = [1 + 0.8 (I$_{A101}$/I$_{R110}$)]$^{-1}$, I$_{A101}$ and I$_{R110}$ represents the integrated intensity of anatase (101 plane) and rutile (110 plane) diffraction peaks, respectively. ^{e} Average crystallite size estimated from the Scherrer's equation, D$_{xrd}$ = kλ/βCosθ, where k is the shape factor (0.9), λ is the X-ray radiation wavelength (0.154 nm), β is the full width at half maxima of the corresponding Bragg angle (θ).

3.3. Microstructural Characterizations

The surface morphology of the TS82C composite was characterized by FESEM analyses, which is presented in Figure S2. The sample before calcination showed small sized nanoparticle formations grown along the cellulose microfibril's surface (Figure S2a,b), whereas after calcination the fibrous assembly disappeared leaving a porous structure where TiO$_2$ nanoparticles with an average size of 20–30 nm are distinctly visible (Figure S2c,d). To further analyze the microscopic particle structure and the arrangement of TiO$_2$/SiO$_2$, TEM measurements were performed on the calcined TS82C composite. Figure 4a shows the low magnification bright field image of TS82C, where two different contrasts are clearly distinguishable. Spherical TiO$_2$ nanoparticles with average size of about 20 nm could be identified from the dark contrast, whereas the lighter contrast corresponds to the amorphous silica support, forming a core-shell-like assembly, which is similar to the microscopic structure of TiO$_2$@SiO$_2$ composite reported by Yuan et al. [55]. So, it can be concluded that the nanocrystalline TiO$_2$ particles are well dispersed into the surrounding mesoporous silica matrix, which also prevents the aggregation of TiO$_2$ particles thus contributing to a higher surface area. Figure 4b,c show the high resolution images, where highly crystalline TiO$_2$ particles with an anatase phase are observed. This is also supported by the formation of high contrast diffraction rings presented in Figure 4d that closely matched the lattice parameters obtained from the XRD result (Table 2). It is also noteworthy to observe some pores with diameter ranging from 3 to 4 nm (Figure 4b) that resemble the mesoporous structure of the TS82C nanocomposite supported by the BET measurements (Table 1), where an average pore diameter of 3.84 nm was observed.

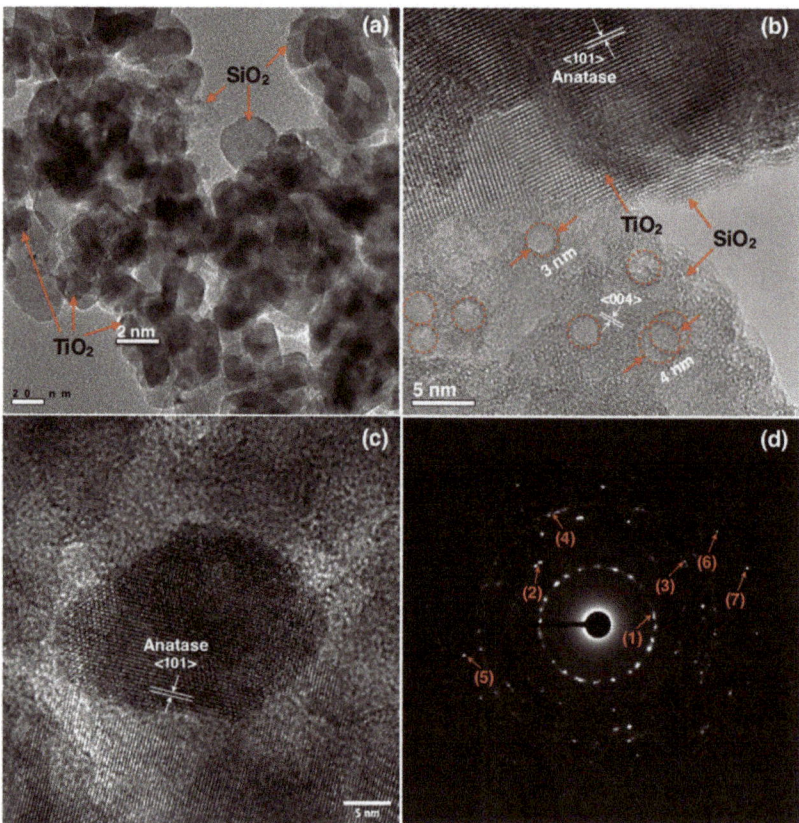

Figure 4. (a–c) TEM images showing morphology and crystallinity of TS82C composite calcined at 550 °C, (a) low magnification, (b,c) high resolution (HRTEM) images, and (d) selected area electron diffraction (SAED) pattern taken from image (c). The lattice spacing of the respective rings numbered in the SAED pattern is summarized in Table 2.

Table 2. Lattice parameters of the nanocrystalline TS82C calculated from SAED (Figure 4d) and XRD pattern (Figure 2a).

Diffraction Ring Number	[a] Lattice Spacing (d, Å)	Miller Indices (hkl)	[b] d_{XRD} (Å)
1	3.50	101	3.51
2	2.34	004	2.36
3	1.87	200	1.89
4	1.68	105	1.69
5	1.47	204	1.47
6	1.34	220	1.33
7	1.26	215	1.27

[a] Calculated using the camera equation, $d = \lambda L/R$, where λ is the wavelength of accelerated electron, L is the camera constant, and R is the radius of the corresponding diffraction ring. [b] Interplanar spacings obtained from 2θ values of the X-ray diffractogram (Figure 2).

3.4. Photocatalytic Activity

The photocatalytic activity of the TiO_2/SiO_2 composite samples was evaluated by the decomposition of three kinds of model pollutants, namely rhodamine B (RhB), Methylene blue (MB), and 4-Nitrophenol (4NP). Although RhB and MB dyes have been extensively studied as model pollutants that are commonly used in textile industries, Buriak et al. [56] suggested the inclusion of phenols as the model pollutant to compare the photodegradation efficiencies as the complete photodegradation of phenolic compounds is quite a difficult task. In fact, it is demonstrated that the reaction rate constant is much lower for 4NP (K value 0.025 min^{-1}) compared to RhB (K value 0.082 min^{-1}) and MB (K value 0.188 min^{-1}). The reaction parameters along with the half-life time are reported in Table 3. Degradation tests following the similar conditions were also performed with commercial P25 TiO_2 powder to compare the results. Figure 5a$_1$,b$_1$,c$_1$ show the degradation kinetics of RhB, MB, and 4NP, respectively. The evolution of the optical absorption spectra under light illumination of each pollutant performed with TS82C and P25 is shown in the (Supporting Information Figures S4–S6). It is clearly observed that in each case, the kinetic rate of dye degradation for TS82C is much higher than P25 TiO_2, which is well pronounced for its very high photocatalytic nature. It is also noteworthy that the degradation efficiency of TS82C is much higher in case of MB with a rate constant of 0.188 min^{-1}, and the lowest calculated half-life time of 3.68 min that might be due to the higher adsorption of MB on TS82C surfaces that triggers the photocatalytic activity by providing more active sites (Figure S5). As shown in Figures S4–S6, it is observed that the adsorption capacity of TS82C is much higher for RhB and MB dye molecules, whereas there is minimal adsorption for the 4NP dye molecules. This can be explained by the electrostatic interaction between the catalyst surface and the dye molecules. Having many hydroxyl groups present on the TS82C surface (evidenced from the FTIR spectra, Figure 1b) we can assume that in a near neutral environment, it would display surface negativity [57], thus attracting and trapping the cationic dye molecules into the porous network and consequently increasing the photodegradation efficiency. On the other hand, under identical experimental conditions, 4NP shows a tendency to exist in an anionic form [57], thus repelling the negatively charged TS82C surface showing poor adsorption and hence a lower photodegradation efficiency.

Table 3. Kinetic parameters of Rhodamine B, Methylene Blue and 4-Nitrophenol obtained from the photodegradation experiments performed with TS82C and P25 TiO_2.

Sample	Rhodamine B			Methylene Blue			4-Nitrophenol		
	[a]K	[b]$t_{1/2}$	[c]R^2	[a]K	[b]$t_{1/2}$	[c]R^2	[a]K	[b]$t_{1/2}$	[c]R^2
TS82C	0.082	8.45	0.98	0.188	3.68	0.97	0.025	27.72	0.88
P25	0.045	15.40	0.99	0.023	30.13	0.99	0.008	86.64	0.97

[a] Reaction rate constant, min^{-1}. [b] Half-life time of the respective organic dyes in minute calculated using the relation $t_{1/2} = \ln 2/K$, where K is the apparent reaction rate constant. [c] Coefficient of determination.

The reusability of the TS82C is reported in Figure 6, where three consecutive runs were performed using RhB, MB, and 4NP dyes. Although RhB and MB showed good consistency (4.8% and 12.2% less efficiency for 2nd and 3rd consecutive runs in case of RhB; 3.2% and 5.3% less efficiency for 2nd and 3rd runs in case of MB), a much lower efficiency was observed for 4NP with an increasing no. of runs (12% and 28% less efficiency for 2nd and 3rd run, respectively). A comparative study on the photocatalytic efficiency of various TiO_2/SiO_2 nanostructures is reported in Table S3. Although photocatalytic efficiency depends on various factors, such as light intensity, illumination, emitting wavelength, catalyst doses, dye concentration etc., Table S3 suggests that TS82C could perform well compared to those reported by the other researchers, particularly in solar photocatalysis. Despite having a lower reaction rate constant for 4NP compared to RhB and MB, TS82C shows a better performance than that reported for 4NP photodegradation [3,58]. Previous works on photocatalytic dye degradation suggest that the entrapment of the semiconductor photocatalysts in amorphous silica matrix enhance the photoactivity [55]. Therefore, we can assume that the mesoporous nature and high surface area of TS82C play the key roles in photodegradation process. This is also evidenced from the much higher photodegradation efficiency of MB, where a higher adsorption was observed (Figure S5). Additionally, the stable anatase crystalline phase in TS82C provides a longer lifetime to the photogenerated electron-hole pairs and favors the adsorption site of the superoxide anions, which also contribute to the photocatalysis. The photogenerated charge carrier separation scenario within the TS82C nanocomposite was verified by the photoluminescence (PL) study and the PL spectra of TS82C and P25 TiO_2, as shown in Figure S7, where a relatively lower intensity of the PL spectra is observed for TS82C compared to P25 TiO_2. The lower PL signal suggests that the photogenerated electrons are trapped within the porous TS82C nanocomposite and transferred to the photocatalytic system, contributing to a higher photocatalytic efficiency.

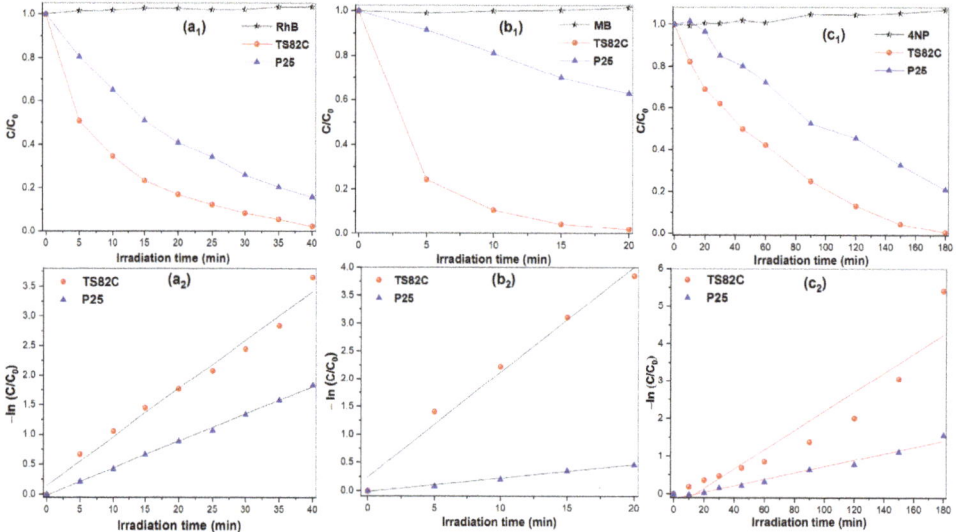

Figure 5. Photo-degradation kinetics of 15 μM aqueous solution of RhB (a_1,a_2), MB (b_1,b_2), and 4NP (c_1,c_2).

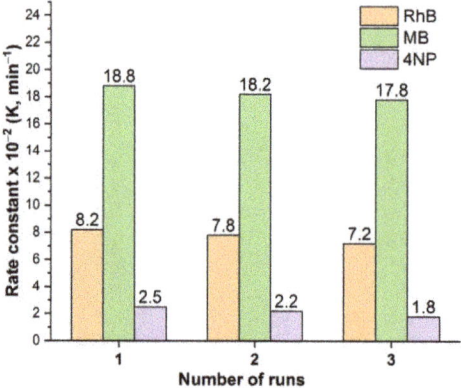

Figure 6. Reusability of TS82C photocatalyst by performing 3 consecutive runs with RhB, MB and 4NP dyes.

4. Conclusions

We have successfully synthesized cellulose and Pluronic F127 templated mesoporous TiO_2/SiO_2 nanocomposites with a higher specific surface area and higher thermal stability of the nanocrystalline anatase phase. An optimum level of silica incorporation (20% by weight) improved the anatase nanocrystalline phase stability, where cellulose/F127 soft-templating increased the mesoporosity while preserving the higher surface area. All the templating media were removed by thermal decomposition at 550 °C, as seen from the TGA/FTIR spectra; thus, a pure TiO_2/SiO_2 nanocomposite was obtained. The highly crystalline spherical TiO_2 particles were dispersed in a mesoporous silica structure as revealed by TEM analyses. The composite sample showed much higher photocatalytic efficiency than TiO_2/SiO_2 without the cellulose templating, and so did the standard reference catalyst P25 TiO_2. The photodegradation efficiency was tested against RhB, MB, and 4NP dyes under solar light irradiation. It could be said that the higher photocatalytic efficiency was achieved due to (i) the higher specific surface area, (ii) the formation of mesoporous structure, and (iii) the highly stable nanocrystalline anatase phase. The composite sample was not only effective against the commonly used dyes (RhB, MB), but also successfully decomposed nitro-aromatic compound (4NP), for which additional co-catalysts are required in the case of photodegradation. Since natural cellulose was used as a second templating, the catalyst powders could be prepared in large scale and eventually extended to synthesize other semiconductor photocatalysts and could find applications in water remediation technology and solar photocatalysis.

Supplementary Materials: The following supporting information can be downloaded at: https://www.mdpi.com/article/10.3390/catal12070770/s1, Table S1. Physicochemical properties of the TiO_2/SiO_2 composites; Figure S1. (a) N_2 adsorption–desorption isotherms and (b) pore size distribution plots of composite powders with different TiO_2/SiO_2 weight ratios; Figure S2. FESEM images of TS82C, (a, b) As prepared composite after drying, (c, d) after calcination at 550 °C; Figure S3. Photocatalytic degradation of RhB dye (15×10^{-6} M) with different TiO_2/SiO_2 photocatalysts; Table S2. Parameters of RhB dye degradation kinetics with different TiO_2/SiO_2 composites; Figure S4. Optical absorption spectra of RhB dye (15×10^{-6} M) with solar light exposure time performed with (a) TS82C and (b) P25 photocatalysts. The adsorption–desorption equilibrium was determined after stirring of 18 h at the dark; Figure S5. Optical absorption spectra of MB dye (15×10^{-6} M) with solar light exposure time performed with (a) TS82C and (b) P25 photocatalysts. The adsorption–desorption equilibrium was determined after stirring of 18 h in dark; Figure S6. Optical absorption spectra of 4NP compound (15×10^{-6} M) with solar light exposure time performed with (a) TS82C and (b) P25 photocatalysts. The adsorption–desorption equilibrium was determined after stirring of 18 h in dark; Figure S7. Photoluminescence (PL) spectra of TS82C nanocomposite (Red line) and P25 TiO_2 (Black line) measured on Horiba JOBIN YVON Fluoromax-P PL spectrophotometer using 365 nm as the excitation wavelength; Table S3. Comparative study of the photocatalytic activity with different TiO_2/SiO_2 photocatalysts [59–66].

Author Contributions: Conceptualization, S.P. and A.L.; methodology, S.P.; software, S.P. and A.T.; validation, S.P.; formal analysis, S.P.; investigation, S.P., A.T. and M.C.; data curation, S.P., A.T. and M.C.; writing—original draft preparation, S.P.; writing—review and editing, S.P. and A.L. All authors have read and agreed to the published version of the manuscript.

Funding: This research received no external funding.

Acknowledgments: The authors thankfully acknowledge Goutam De for providing the GIXRD facility, Fabio Marzo for performing the FESEM measurements, and Donato Cannoletta for recording the wide angle XRD spectra.

Conflicts of Interest: The authors declare no conflict of interest.

References

1. Hoffmann, M.R.; Martin, S.T.; Choi, W.; Bahnemann, D.W. Environmental Applications of Semiconductor Photocatalysis. *Chem. Rev.* **1995**, *95*, 69–96. [CrossRef]
2. Joseph, C.G.; Taufiq-Yap, Y.H.; Letshmanan, E.; Vijayan, V. Heterogeneous Photocatalytic Chlorination of Methylene Blue Using a Newly Synthesized TiO_2-SiO_2 Photocatalyst. *Catalysts* **2022**, *12*, 156. [CrossRef]
3. San, N.; Hatipoğlu, A.; Koçtürk, G.; Çınar, Z. Photocatalytic degradation of 4-nitrophenol in aqueous TiO_2 suspensions: Theoretical prediction of the intermediates. *J. Photochem. Photobiol. Chem.* **2002**, *146*, 189–197. [CrossRef]
4. Hashimoto, K.; Irie, H.; Fujishima, A. TiO_2 photocatalysis: A historical overview and future prospects. *Jpn. J. Appl. Phys.* **2005**, *44*, 8269. [CrossRef]
5. Chen, H.; Nanayakkara, C.E.; Grassian, V.H. Titanium Dioxide Photocatalysis in Atmospheric Chemistry. *Chem. Rev.* **2012**, *112*, 5919–5948. [CrossRef]
6. Dimitroula, H.; Daskalaki, V.M.; Frontistis, Z.; Kondarides, D.I.; Panagiotopoulou, P.; Xekoukoulotakis, N.P.; Mantzavinos, D. Solar photocatalysis for the abatement of emerging micro-contaminants in wastewater: Synthesis, characterization and testing of various TiO_2 samples. *Appl. Catal. B Environ.* **2012**, *117*, 283–291. [CrossRef]
7. He, Y.; Sutton, N.B.; Rijnaarts, H.H.H.; Langenhoff, A.A.M. Degradation of pharmaceuticals in wastewater using immobilized TiO_2 photocatalysis under simulated solar irradiation. *Appl. Catal. B Environ.* **2016**, *182*, 132–141. [CrossRef]
8. McCullagh, C.; Skillen, N.; Adams, M.; Robertson, P.K.J. Photocatalytic reactors for environmental remediation: A review. *J. Chem. Technol. Biotechnol.* **2011**, *86*, 1002–1017. [CrossRef]
9. Mills, A.; O'Rourke, C.; Moore, K. Powder semiconductor photocatalysis in aqueous solution: An overview of kinetics-based reaction mechanisms. *J. Photochem. Photobiol. Chem.* **2015**, *310*, 66–105. [CrossRef]
10. Maeda, K.; Domen, K. New Non-Oxide Photocatalysts Designed for Overall Water Splitting under Visible Light. *J. Phys. Chem. C* **2007**, *111*, 7851–7861. [CrossRef]
11. Fujishima, A.; Zhang, X.; Tryk, D.A. TiO_2 photocatalysis and related surface phenomena. *Surf. Sci. Rep.* **2008**, *63*, 515–582. [CrossRef]
12. Herrmann, J.-M. Photocatalysis fundamentals revisited to avoid several misconceptions. *Appl. Catal. B Environ.* **2010**, *99*, 461–468. [CrossRef]
13. Etacheri, V.; Seery, M.K.; Hinder, S.J.; Pillai, S.C. Oxygen Rich Titania: A Dopant Free, High Temperature Stable, and Visible-Light Active Anatase Photocatalyst. *Adv. Funct. Mater.* **2011**, *21*, 3744–3752. [CrossRef]
14. Joo, J.B.; Zhang, Q.; Lee, I.; Dahl, M.; Zaera, F.; Yin, Y. Mesoporous Anatase Titania Hollow Nanostructures though Silica-Protected Calcination. *Adv. Funct. Mater.* **2012**, *22*, 166–174. [CrossRef]
15. Yang, P.; Zhao, D.; Margolese, D.I.; Chmelka, B.F.; Stucky, G.D. Block Copolymer Templating Syntheses of Mesoporous Metal Oxides with Large Ordering Lengths and Semicrystalline Framework. *Chem. Mater.* **1999**, *11*, 2813–2826. [CrossRef]
16. Zhao, D.; Huo, Q.; Feng, J.; Chmelka, B.F.; Stucky, G.D. Nonionic Triblock and Star Diblock Copolymer and Oligomeric Surfactant Syntheses of Highly Ordered, Hydrothermally Stable, Mesoporous Silica Structures. *J. Am. Chem. Soc.* **1998**, *120*, 6024–6036. [CrossRef]
17. Araujo, P.Z.; Luca, V.; Bozzano, P.B.; Bianchi, H.L.; Soler-Illia, G.J.d.Á.A.; Blesa, M.A. Aerosol-Assisted Production of Mesoporous Titania Microspheres with Enhanced Photocatalytic Activity: The Basis of an Improved Process. *ACS Appl. Mater. Interfaces* **2010**, *2*, 1663–1673. [CrossRef]
18. Ismail, A.A.; Bahnemann, D.W. Mesoporous titania photocatalysts: Preparation, characterization and reaction mechanisms. *J. Mater. Chem.* **2011**, *21*, 11686–11707. [CrossRef]
19. Choi, S.Y.; Mamak, M.; Coombs, N.; Chopra, N.; Ozin, G.A. Thermally Stable Two-Dimensional Hexagonal Mesoporous Nanocrystalline Anatase, Meso-nc-TiO_2: Bulk and Crack-Free Thin Film Morphologies. *Adv. Funct. Mater.* **2004**, *14*, 335–344. [CrossRef]
20. Mahoney, L.; Koodali, R.T. Versatility of Evaporation-Induced Self-Assembly (EISA) Method for Preparation of Mesoporous TiO_2 for Energy and Environmental Applications. *Materials* **2014**, *7*, 2697–2746. [CrossRef]
21. Chen, L.; Yao, B.; Cao, Y.; Fan, K. Synthesis of Well-Ordered Mesoporous Titania with Tunable Phase Content and High Photoactivity. *J. Phys. Chem. C* **2007**, *111*, 11849–11853. [CrossRef]
22. Wang, W.; Nguyen, D.; Long, H.; Liu, G.; Li, S.; Yue, X.; Ru, H. High temperature and water-based evaporation-induced self-assembly approach for facile and rapid synthesis of nanocrystalline mesoporous TiO_2. *J. Mater. Chem. A* **2014**, *2*, 15912–15920. [CrossRef]
23. Hirano, M.; Ota, K.; Iwata, H. Direct Formation of Anatase (TiO_2)/Silica (SiO_2) Composite Nanoparticles with High Phase Stability of 1300 °C from Acidic Solution by Hydrolysis under Hydrothermal Condition. *Chem. Mater.* **2004**, *16*, 3725–3732. [CrossRef]
24. Dong, W.; Sun, Y.; Lee, C.W.; Hua, W.; Lu, X.; Shi, Y.; Zhang, S.; Chen, J.; Zhao, D. Controllable and Repeatable Synthesis of Thermally Stable Anatase Nanocrystal−Silica Composites with Highly Ordered Hexagonal Mesostructures. *J. Am. Chem. Soc.* **2007**, *129*, 13894–13904. [CrossRef]
25. Chen, S.-Y.; Tang, C.-Y.; Lee, J.-F.; Jang, L.-Y.; Tatsumi, T.; Cheng, S. Effect of calcination on the structure and catalytic activities of titanium incorporated SBA-15. *J. Mater. Chem.* **2011**, *21*, 2255–2265. [CrossRef]

26. Pierpaoli, M.; Zheng, X.; Bondarenko, V.; Fava, G.; Ruello, M.L. Paving the Way for A Sustainable and Efficient SiO_2/TiO_2 Photocatalytic Composite. *Environments* **2019**, *6*, 87. [CrossRef]
27. Eddy, D.R.; Ishmah, S.N.; Permana, M.D.; Firdaus, M.L. Synthesis of Titanium Dioxide/Silicon Dioxide from Beach Sand as Photocatalyst for Cr and Pb Remediation. *Catalysts* **2020**, *10*, 1248. [CrossRef]
28. Temerov, F.; Haapanen, J.; Mäkelä, J.M.; Saarinen, J.J. Photocatalytic Activity of Multicompound TiO_2/SiO_2 Nanoparticles. *Inorganics* **2021**, *9*, 21. [CrossRef]
29. Mohamed, M.A.; Salleh, W.W.; Jaafar, J.; Ismail, A.F.; Mutalib, M.A.; Sani, N.; Asri, S.M.; Ong, C. Physicochemical characteristic of regenerated cellulose/N-doped TiO_2 nanocomposite membrane fabricated from recycled newspaper with photocatalytic activity under UV and visible light irradiation. *Chem. Eng. J.* **2016**, *284*, 202–215. [CrossRef]
30. Zeng, J.; Liu, S.; Cai, J.; Zhang, L. TiO_2 Immobilized in Cellulose Matrix for Photocatalytic Degradation of Phenol under Weak UV Light Irradiation. *J. Phys. Chem. C* **2010**, *114*, 7806–7811. [CrossRef]
31. Liu, X.; Gu, Y.; Huang, J. Hierarchical, Titania-Coated, Carbon Nanofibrous Material Derived from a Natural Cellulosic Substance. *Chem.—Eur. J.* **2010**, *16*, 7730–7740. [CrossRef] [PubMed]
32. Postnova, I.; Kozlova, E.; Cherepanova, S.; Tsybulya, S.; Rempel, A.; Shchipunov, Y. Titania synthesized through regulated mineralization of cellulose and its photocatalytic activity. *RSC Adv.* **2015**, *5*, 8544–8551. [CrossRef]
33. Plumejeau, S.; Rivallin, M.; Brosillon, S.; Ayral, A.; Boury, B. M-Doped TiO_2 and TiO_2–M_xO_y Mixed Oxides (M = V, Bi, W) by Reactive Mineralization of Cellulose—Evaluation of Their Photocatalytic Activity. *Eur. J. Inorg. Chem.* **2016**, *2016*, 1200–1205. [CrossRef]
34. Liciulli, A.; Nisi, R.; Pal, S.; Laera, A.M.; Creti, P.; Chiechi, A. Photo-oxidation of ethylene over mesoporous TiO_2/SiO_2 catalysts. *Adv. Hortic. Sci.* **2016**, *30*, 75–80. [CrossRef]
35. de Chiara, M.; Pal, S.; Licciulli, A.; Amodio, M.; Colelli, G. Photocatalytic degradation of ethylene on mesoporous TiO_2/SiO_2 nanocomposites: Effects on the ripening of mature green tomatoes. *Biosyst. Eng.* **2015**, *132*, 61–70. [CrossRef]
36. Hongo, T.; Yamazaki, A. Thermal influence on the structure and photocatalytic activity of mesoporous titania consisting of $TiO_2(B)$. *Microporous Mesoporous Mater.* **2011**, *142*, 316–321. [CrossRef]
37. Crepaldi, E.L.; Soler-Illia, G.J.D.A.A.; Bouchara, A.; Grosso, D.; Durand, D.; Sanchez, C. Controlled Formation of Highly Ordered Cubic and Hexagonal Mesoporous Nanocrystalline Yttria–Zirconia and Ceria–Zirconia Thin Films Exhibiting High Thermal Stability. *Angew. Chem. Int. Ed.* **2003**, *42*, 347–351. [CrossRef]
38. Luo, Y.; Xu, J.; Huang, J. Hierarchical nanofibrous anatase-titania–cellulose composite and its photocatalytic property. *CrystEngComm* **2014**, *16*, 464–471. [CrossRef]
39. Padmanabhan, S.K.; Pal, S.; Haq, E.U.; Licciulli, A. Nanocrystalline TiO_2–diatomite composite catalysts: Effect of crystallization on the photocatalytic degradation of rhodamine B. *Appl. Catal. Gen.* **2014**, *485*, 157–162. [CrossRef]
40. Cunha, A.G.; Freire, C.S.R.; Silvestre, A.J.D.; Neto, C.P.; Gandini, A.; Orblin, E.; Fardim, P. Highly Hydrophobic Biopolymers Prepared by the Surface Pentafluorobenzoylation of Cellulose Substrates. *Biomacromolecules* **2007**, *8*, 1347–1352. [CrossRef]
41. Ciolacu, D.; Ciolacu, F.; Popa, V.I. Amorphous cellulose—Structure and characterization. *Cellul. Chem. Technol.* **2011**, *45*, 13.
42. Carboni, D.; Marongiu, D.; Rassu, P.; Pinna, A.; Amenitsch, H.; Casula, M.; Marcelli, A.; Cibin, G.; Falcaro, P.; Malfatti, L.; et al. Enhanced Photocatalytic Activity in Low-Temperature Processed Titania Mesoporous Films. *J. Phys. Chem. C* **2014**, *118*, 12000–12009. [CrossRef]
43. Saha, J.; De, G. Highly ordered cubic mesoporous electrospun SiO_2 nanofibers. *Chem. Commun.* **2013**, *49*, 6322–6324. [CrossRef]
44. Chattopadhyay, S.; Saha, J.; De, G. Electrospun anatase TiO_2 nanofibers with ordered mesoporosity. *J. Mater. Chem. A* **2014**, *2*, 19029–19035. [CrossRef]
45. Pal, S.; Laera, A.M.; Licciulli, A.; Catalano, M.; Taurino, A. Biphase TiO_2 Microspheres with Enhanced Photocatalytic Activity. *Ind. Eng. Chem. Res.* **2014**, *53*, 7931–7938. [CrossRef]
46. Parra, R.; Góes, M.S.; Castro, M.S.; Longo, E.; Bueno, P.R.; Varela, J.A. Reaction Pathway to the Synthesis of Anatase via the Chemical Modification of Titanium Isopropoxide with Acetic Acid. *Chem. Mater.* **2008**, *20*, 143–150. [CrossRef]
47. De Ceglie, C.; Pal, S.; Murgolo, S.; Licciulli, A.; Mascolo, G. Investigation of Photocatalysis by Mesoporous Titanium Dioxide Supported on Glass Fibers as an Integrated Technology for Water Remediation. *Catalysts* **2022**, *12*, 41. [CrossRef]
48. Yang, P.; Zhao, D.; Margolese, D.I.; Chmelka, B.F.; Stucky, G.D. Generalized syntheses of large-pore mesoporous metal oxides with semicrystalline frameworks. *Nature* **1998**, *396*, 152–155. [CrossRef]
49. Zhao, D.; Yang, P.; Melosh, N.; Feng, J.; Chmelka, B.F.; Stucky, G.D. Continuous Mesoporous Silica Films with Highly Ordered Large Pore Structures. *Adv. Mater.* **1998**, *10*, 1380–1385. [CrossRef]
50. Zhang, J.; Li, M.; Feng, Z.; Chen, J.; Li, C. UV Raman Spectroscopic Study on TiO_2. I. Phase Transformation at the Surface and in the Bulk. *J. Phys. Chem. B* **2006**, *110*, 927–935. [CrossRef]
51. Sing, K.S.W. Reporting Physisorption Data for Gas/Solid Systems with Special Reference to the Determination of Surface Area and Porosity. *Pure Appl. Chem.* **1985**, *57*, 603–619. [CrossRef]
52. Chen, D.; Huang, F.; Cheng, Y.-B.; Caruso, R.A. Mesoporous Anatase TiO_2 Beads with High Surface Areas and Controllable Pore Sizes: A Superior Candidate for High-Performance Dye-Sensitized Solar Cells. *Adv. Mater.* **2009**, *21*, 2206–2210. [CrossRef]
53. Luo, Z.; Poyraz, A.S.; Kuo, C.-H.; Miao, R.; Meng, Y.; Chen, S.-Y.; Jiang, T.; Wenos, C.; Suib, S.L. Crystalline Mixed Phase (Anatase/Rutile) Mesoporous Titanium Dioxides for Visible Light Photocatalytic Activity. *Chem. Mater.* **2015**, *27*, 6–17. [CrossRef]

54. Ko, Y.G.; Lee, H.J.; Kim, J.Y.; Choi, U.S. Hierarchically Porous Aminosilica Monolith as a CO_2 Adsorbent. *ACS Appl. Mater. Interfaces* **2014**, *6*, 12988–12996. [CrossRef] [PubMed]
55. Yuan, L.; Han, C.; Pagliaro, M.; Xu, Y.-J. Origin of Enhancing the Photocatalytic Performance of TiO_2 for Artificial Photoreduction of CO_2 through a SiO_2 Coating Strategy. *J. Phys. Chem. C* **2016**, *120*, 265–273. [CrossRef]
56. Buriak, J.M.; Kamat, P.V.; Schanze, K.S. Best Practices for Reporting on Heterogeneous Photocatalysis. *ACS Appl. Mater. Interfaces* **2014**, *6*, 11815–11816. [CrossRef]
57. Deng, F.; Liu, Y.; Luo, X.; Wu, S.; Luo, S.; Au, C.; Qi, R. Sol-hydrothermal synthesis of inorganic-framework molecularly imprinted TiO_2/SiO_2 nanocomposite and its preferential photocatalytic degradation towards target contaminant. *J. Hazard. Mater.* **2014**, *278*, 108–115. [CrossRef]
58. Rezaei-Vahidian, H.; Zarei, A.R.; Soleymani, A.R. Degradation of nitro-aromatic explosives using recyclable magnetic photocatalyst: Catalyst synthesis and process optimization. *J. Hazard. Mater.* **2017**, *325*, 310–318. [CrossRef]
59. Kim, Y.N.; Shao, G.N.; Jeon, S.J.; Imran, S.M.; Sarawade, P.B.; Kim, H.T. Gel Synthesis of Sodium Silicate and Titanium Oxychloride Based TiO_2–SiO_2 Aerogels and Their Photocatalytic Property under UV Irradiation. *Chem. Eng. J.* **2013**, *231*, 502–511. [CrossRef]
60. Mahanta, U.; Khandelwal, M.; Deshpande, A.S. TiO_2@SiO_2 Nanoparticles for Methylene Blue Removal and Photocatalytic Degradation under Natural Sunlight and Low-Power UV Light. *Appl. Surf. Sci.* **2022**, *576*, 151745. [CrossRef]
61. Bao, Y.; Guo, R.; Gao, M.; Kang, Q.; Ma, J. Morphology Control of 3D Hierarchical Urchin-like Hollow SiO_2@TiO_2 Spheres for Photocatalytic Degradation: Influence of Calcination Temperature. *J. Alloys Compd.* **2021**, *853*, 157202. [CrossRef]
62. Ferreira-Neto, E.P.; Ullah, S.; Simões, M.B.; Perissinotto, A.P.; de Vicente, F.S.; Noeske, P.-L.M.; Ribeiro, S.J.L.; Rodrigues-Filho, U.P. Solvent-Controlled Deposition of Titania on Silica Spheres for the Preparation of SiO_2@TiO_2 Core@shell Nanoparticles with Enhanced Photocatalytic Activity. *Colloids Surf. A Physicochem. Eng. Asp.* **2019**, *570*, 293–305. [CrossRef]
63. Guo, N.; Liang, Y.; Lan, S.; Liu, L.; Ji, G.; Gan, S.; Zou, H.; Xu, X. Uniform TiO_2–SiO_2 Hollow Nanospheres: Synthesis, Characterization and Enhanced Adsorption–Photodegradation of Azo Dyes and Phenol. *Appl. Surf. Sci.* **2014**, *305*, 562–574. [CrossRef]
64. Palhares, H.G.; Nunes, E.H.M.; Houmard, M. Heat Treatment as a Key Factor for Enhancing the Photodegradation Performance of Hydrothermally-Treated Sol–Gel TiO_2–SiO_2 Nanocomposites. *J. Sol-Gel Sci. Technol.* **2021**, *99*, 188–197. [CrossRef]
65. Bellardita, M.; Addamo, M.; Di Paola, A.; Marcì, G.; Palmisano, L.; Cassar, L.; Borsa, M. Photocatalytic Activity of TiO_2/SiO_2 Systems. *J. Hazard. Mater.* **2010**, *174*, 707–713. [CrossRef]
66. Liao, S.; Lin, L.; Huang, J.; Jing, X.; Chen, S.; Li, Q. Microorganism-Templated Nanoarchitectonics of Hollow TiO_2-SiO_2 Microspheres with Enhanced Photocatalytic Activity for Degradation of Methyl Orange. *Nanomaterials* **2022**, *12*, 1606. [CrossRef]

Article

Toluene Decomposition in Plasma–Catalytic Systems with Nickel Catalysts on CaO-Al$_2$O$_3$ Carrier

Joanna Woroszył-Wojno *, Michał Młotek, Bogdan Ulejczyk and Krzysztof Krawczyk

Faculty of Chemistry, Warsaw University of Technology, 3 Noakowskiego Street, 00-664 Warsaw, Poland; mmlotek@ch.pw.edu.pl (M.M.); bulejczyk@ch.pw.edu.pl (B.U.); kraw@ch.pw.edu.pl (K.K.)
* Correspondence: joanna.woroszyl.dokt@pw.edu.pl

Abstract: The decomposition of toluene as a tar imitator in a gas composition similar to the gas after biomass pyrolysis was studied in a plasma–catalytic system. Nickel catalysts and the plasma from gliding arc discharge under atmospheric pressure were used. The effect of the catalyst bed, discharge power, initial toluene, and hydrogen concentration on C_7H_8 decomposition, calorific value, and unit energy consumption were studied. The gas flow rate was 1000 NL/h, while the inlet gas composition (molar ratio) was CO (0.13), CO_2 (0.15), H_2 (0.28–0.38), and N_2 (0.34–0.44). The study was conducted using an initial toluene concentration in the range of 2000–4500 ppm and a discharge power of 1500–2000 W. In plasma–catalytic systems, the following catalysts were compared: NiO/Al$_2$O$_3$, NiO/(CaO-Al$_2$O$_3$), and Ni/(CaO-Al$_2$O$_3$). The decomposition of toluene increased with its initial concentration. An increase in hydrogen concentration resulted in higher activity of the Ni/(CaO-Al$_2$O$_3$) catalysts. The gas composition did not change by more than 10% during the process. Trace amounts of C2 hydrocarbons were observed. The conversion of C_7H_8 was up to 85% when NiO/(CaO-Al$_2$O$_3$) was used. The products of the toluene decomposition reactions were not adsorbed onto its surface. The calorific value was not changed during the process and was higher than required for turbines and engines in every system studied.

Keywords: gliding discharge; plasma–catalytic system; tar decomposition; nickel catalyst

Citation: Woroszył-Wojno, J.; Młotek, M.; Ulejczyk, B.; Krawczyk, K. Toluene Decomposition in Plasma–Catalytic Systems with Nickel Catalysts on CaO-Al$_2$O$_3$ Carrier. *Catalysts* 2022, 12, 635. https://doi.org/10.3390/catal12060635

Academic Editor: Jean-François Lamonier

Received: 16 May 2022
Accepted: 7 June 2022
Published: 10 June 2022

Publisher's Note: MDPI stays neutral with regard to jurisdictional claims in published maps and institutional affiliations.

Copyright: © 2022 by the authors. Licensee MDPI, Basel, Switzerland. This article is an open access article distributed under the terms and conditions of the Creative Commons Attribution (CC BY) license (https://creativecommons.org/licenses/by/4.0/).

1. Introduction

The gas produced during the gasification and pyrolysis of biomass contains a significant amount of pollutants, mainly aromatic hydrocarbons [1]. Depending on the type of biomass, the amount of tars varies from 5 to 100 g/Nm^3 [2,3]. As a result, up to 15% of the effective energy of biomass is lost, which is against the principles of a clean energy source [4]. In addition, unpurified biogas cannot be used as a fuel for engines (limit 50–100 mg/Nm^3 of tar) or turbines (5 mg/Nm^3) because the tar concentration is too high [5–7].

Many methods have been developed to remove tars using catalysts, plasma, adsorption, or filtration, depending on the requirements that the gas after pyrolysis or gasification must meet [8–12]. In catalytical methods, nickel has been shown to be the most efficient active phase while set on oxide carriers, such as Al$_2$O$_3$. The main disadvantage of using nickel catalysts has been deactivation of the catalyst, mostly due to the formation of carbon deposits on its surface [13–16].

Plasma and plasma–catalytic systems for toluene decomposition have been successfully conducted in many scientific groups [17–19]. The use of both plasma and catalysts deserves special attention similar to the tar decomposition method due to being more efficient compared to the separate use of catalyst and plasma. However, most studies have been conducted in small-scale reactors with low gas flow ratio and discharge power.

A new Ni$_3$Al catalyst in the form of a honeycomb as well as commercial nickel catalysts designed for carbon oxide methanation (RANG-19PR) and methane water shift (G-0117) have been successfully used in plasma–catalytic systems for the decomposition of C_7H_8, using a flow rate of 1 Nm^3/h and an inlet gas composition similar to that produced in biomass gasification [20–23]. However, these catalysts could be deactivated by carbon deposits formed during the process. Reducing the amount of active phase on the catalyst's bed could result in less soot formation. It was found that a high toluene conversion rate could be achieved with 10% nickel on Al$_2$O$_3$, but soot formation remained a problem; therefore, further studies were needed [24]. The addition of trace amounts of calcium to a catalyst used for catalytic reforming resulted in an increase in its resistance and activity by reducing the formation of carbon deposits [21,25–27]. New nickel catalysts were prepared on a CaO-Al$_2$O$_3$ bed to study the effect of calcium oxide addition in the decomposition of toluene in the plasma–catalytic system. Furthermore, catalysts' activity, the reaction products deposited on the catalyst bed, the calorific value of the outlet gas, the unit energy consumption, and the initial concentrations of C$_7$H$_8$ and hydrogen in plasma–catalytic systems were studied.

2. Results and Discussion

In the plasma–catalytic system, higher toluene conversion ($x_{C_7H_8}$) was observed than in systems without catalysts (plasma only). The highest $x_{C_7H_8}$ was observed for the coupled system with the NiO/(CaO-Al$_2$O$_3$) catalyst—85%. In the plasma–catalytic system with Ni/(CaO-Al$_2$O$_3$) and NiO/Al$_2$O$_3$ catalysts [24], the highest results were 77% and 82%, respectively. Without a catalyst, up to 68% of toluene was decomposed [21].

The toluene conversions obtained were lower than those obtained with a commercial nickel catalyst (G-0117, manufactured by INS Pulawy, Poland) used in the plasma–catalytic system (conversion—99%). It could be a consequence of lower temperature during the process and lower concentration of active phase on the catalyst [22,23,28].

The trace addition of calcium oxide onto the catalyst surface prevented the sintering of the catalyst and changed the interaction between the active phase and the catalyst carrier. This resulted in an increase in toluene conversion [26,29]. Studies on the effect of Ca addition on the steam reforming of ethanol showed that the addition of a trace amount of calcium to the nickel catalyst on Al$_2$O$_3$ support increased coke formation on its surface. However, due to the rapid formation, it was less stable and amorphous, making it easier to oxidize than the graphitic carbon formed on the nickel catalyst on Al$_2$O$_3$ without Ca addition [30].

A methanation reaction of carbon oxide occurred in all systems studied. A higher methane concentration was observed in a gas composition with a higher hydrogen concentration (series B) because the reaction rate at the catalyst was higher. More H$_2$ in the initial gas lowered the toluene conversion in plasma–catalytic systems with NiO/(CaO-Al$_2$O$_3$) and NiO/Al$_2$O$_3$ [24] but increased in the system with the Ni/(CaO-Al$_2$O$_3$) catalyst. In series B, more hydrogen radicals were present during the process, which could be used not only for the methanation reaction but also for the hydrogenation of toluene radical, resulting in lower C$_7$H$_8$ conversion rates in the system with NiO/(CaO-Al$_2$O$_3$). The amount of CH$_4$ produced in the plasma–catalytic system with the use of Ni/(CaO-Al$_2$O$_3$) catalyst reached 600 ppm (Figure 1), which was significantly higher than that produced with NiO/(CaO-Al$_2$O$_3$)—70 ppm—or Ni/Al$_2$O$_3$—50 ppm [24]. The addition of CaO to the catalyst carrier resulted in an increased methane concentration in the outlet gas.

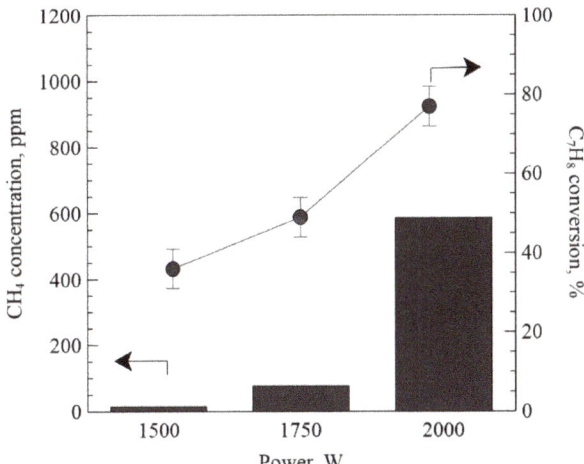

Figure 1. Methane formation and toluene conversion from the molar fraction of hydrogen in inlet gas and the discharge power. Initial toluene concentration 2000 ppm, catalyst Ni/(CaO-Al$_2$O$_3$), series B.

The highest conversions were observed with the highest initial toluene concentration (4500 ppm) in each system studied, as this increased the reaction rate on the catalyst surface. In contrast to previous studies, a slightly higher decomposition of toluene was observed when catalysts with oxidized nickel were used [22,23]. This could be related to the fact that byproducts of toluene decomposition were adsorbed on Ni/(CaO-Al$_2$O$_3$), which limited the activity of the catalyst due to blocked access to the active sites.

Toluene conversion rates were stable for NiO/(CaO-Al$_2$O$_3$) and NiO/Al$_2$O$_3$ catalysts and an increase in power did not increase $x_{C_7H_8}$ because the catalytical process affected toluene decomposition to a greater extent than the plasma one (Figure 2). While the Ni/(CaO-Al$_2$O$_3$) catalyst was used, conversion rates increased with increasing discharge power because the process was more dependent on plasma, and more active radicals could react with toluene.

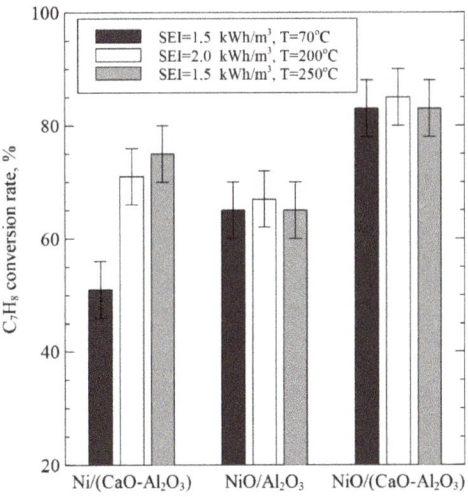

Figure 2. Effect of *SEI* and temperature on toluene conversion. Initial toluene concentration 4500 ppm, series A.

In previous studies, a reduction of NiO to Ni was observed when the G-0117 [23] catalyst was used. This had a positive effect since Ni was more active than NiO in the decomposition of tar. However, in the case of this study, this might have had a negative influence. The reason was the different Ni/Ca ratio in the catalyst bed. A very low addition of calcium (Ca/Ni below 0.2) increased the resistance of the Ni/CaO-Al$_2$O$_3$ catalyst to sintering and the formation of carbon deposits, while a higher ratio resulted in greater soot formation. This was due to the growth of Ni crystallites and the coverage of the catalyst surface with Ca, which hindered the interaction of Ni with the catalyst bed [26,30]. The optimal Ca/Ni ratio for the NiO/CaO-Al$_2$O$_3$ catalyst might be different from that of the Ni/CaO-Al$_2$O$_3$ catalyst, which would explain why the addition of calcium did not decrease the toluene conversion rate in the plasma–catalytic system with NiO/CaO-Al$_2$O$_3$ [29].

In previous studies with plasma–catalytic systems and nickel catalysts, it was observed that turning off the plasma discharge resulted in a rapid decrease in toluene conversion [21]. Therefore, the process of toluene decomposition required both the presence of plasma and temperature to achieve high C$_7$H$_8$ conversion rates.

Energy efficiency (*EE*) calculations were used to compare the results obtained with those of other studies that used gliding arc discharge (GA) and a nickel catalyst in a coupled plasma–catalytic system for tar decomposition. At an initial toluene concentration of 4500 ppm (18.5 g/m^3), *EE* values for NiO/Al$_2$O$_3$, NiO/(CaO-Al$_2$O$_3$), and Ni/(CaO-Al$_2$O$_3$) catalysts in plasma–catalytic systems were up to 7.5 g/kWh, 9.0 g/kWh and 8.9 g/kWh, respectively. It decreased with the increase in *SEI*, while the conversion rates of toluene remained stable (Figure 3).

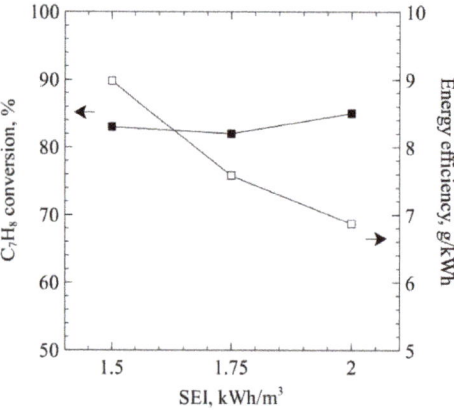

Figure 3. Effect of *SEI* on toluene conversion and energy efficiency for 4500 ppm initial C$_7$H$_8$ concentration in the plasma–catalytic system with NiO/(CaO-Al$_2$O$_3$) catalyst. Series A.

Tar imitator conversion rates obtained in this study were lower than those reported in other groups in which as much as even 95.7% of toluene and 83.4% of naphthalene were decomposed when the Ni-Co/γ-Al$_2$O$_3$ catalyst was used in the plasma–catalytic with gliding arc plasma system [31]. Other studies also reported a very high conversion of toluene up to 95.2% when the Ni/γ-Al$_2$O$_3$ catalyst with rotating gliding arc discharge was used [32]. The energy efficiency values obtained in the studies were in a similar range to those reported in the literature, i.e., 6.7 g/kWh [20] and 3.6 g/kWh [33]. However, when GA discharge plasma was used, much higher *EE* values were obtained, up to 46.3 g/kWh [34]. The difference between *EE* and toluene conversion rates reported in the literature and in this study could be due to the lower gas flow rate (up to 0.54 Nm^3/h) [32], which allowed for longer residence time and contact of excited radicals with toluene particles.

The calorific values of the gas after the process were similar to the initial calorific values. They were as high as 5.8 MJ/m^3 and 7.3 MJ/m^3 for series A and B, respectively, in the system with the NiO/(CaO-Al$_2$O$_3$) catalyst (Figure 4).

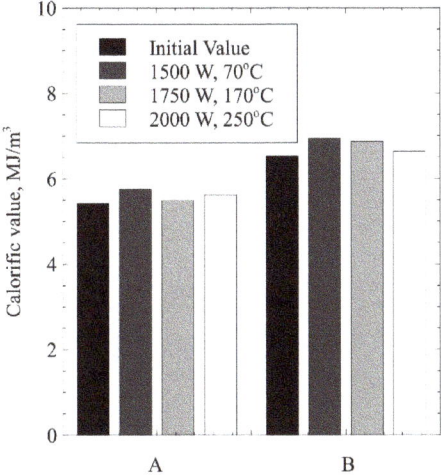

Figure 4. Effect of temperature and discharge power on calorific value for 4500 ppm initial C$_7$H$_8$ concentration in the plasma–catalytic system with NiO/(CaO-Al$_2$O$_3$) catalyst.

3. Experimental

3.1. Catalysts Preparation

Catalysts NiO/Al$_2$O$_3$, NiO/(CaO-Al$_2$O$_3$), and Ni/(CaO-Al$_2$O$_3$) with 10 wt% active phase on Al$_2$O$_3$ and on the commercial catalysts support G-2117-7H/C (CaO-Al$_2$O$_3$), manufactured by INS Pulawy, Poland, had a specific surface area not exceeding 10 m^2/g (Table 1). In previous studies, catalysts on the G-117 bed were successfully used and the optimal concentration of the active phase was identified to be 10 wt% [21–24]. Catalysts were prepared by the impregnation method using Ni(NO$_3$)$_2$·6H$_2$O, which was deposited on the catalyst support and dried at 90 °C for 3 h. Afterward, they were calcinated at 500 °C for 5 h. A catalyst with metallic active phase was reduced at 400 °C for 14 h under 4 Nl/h hydrogen flow rate.

Table 1. The specific surface area before and after the process [24].

S$_{BET}$ (m^2/g)					
NiO/(CaO-Al$_2$O$_3$)		Ni/(CaO-Al$_2$O$_3$)		NiO/Al$_2$O$_3$ [24]	
Before	After	Before	After	Before	After
4.2	3.6	2.7	2.5	8.6	8.0

3.2. Methods

Two gas compositions similar to the gas composition after biomass gasification were used (series A and B) in the plasma–catalytic systems. The CO and CO$_2$ amounts were the same for both series (Table 2). The total inlet gas flow was 1000 Nl/h, and the initial toluene concentrations were 2000, 3000, and 4500 ppm. The gliding arc discharge power was in the range of 1500–2000 W and was measured with an energy meter, Schrack MGDIZ065. The catalyst bed was placed over the end of the electrodes. A detailed description of the reactor was given in a previous article [23].

Table 2. Inlet gas composition.

Composition	H$_2$	N$_2$	CO	CO$_2$
A	0.28	0.44	0.13	0.15
B	0.38	0.34	0.13	0.15

Gas samples were taken for three increasing discharge power; afterwards, the discharge power was reduced to the initial value (1500 W) to study the effect of the catalyst temperature on toluene decomposition. The results obtained were compared to those of plasma-only and plasma–catalytic systems with commercial nickel catalysts [21–23].

After the process, the catalysts were taken out from the reactor and rinsed with acetone. The solutions obtained were analyzed by Thermo-Scientific ISQ mass spectrometer. The gas composition before and after the process was analyzed using gas chromatograph Agilent 6890 N with ShinCarbon column and TCD and FID detectors.

Toluene conversion rate was calculated using the following equation [23]:

$$x = \frac{Co - C}{Co} \tag{1}$$

x—toluene conversion,
Co—initial toluene concentration (g/m^3), and
C—toluene concentration on outlet gas (g/m^3).

For the calculation of the calorific value, the following equation was used [21,23]:

$$W = \frac{Q_{p\,H_2} \cdot n_{H_2} + Q_{p\,CO} \cdot n_{CO} + Q_{p\,CH_4} \cdot n_{CH_4} + Q_{p\,C_2H_2} \cdot n_{C_2H_2} + Q_{p\,C_2H_4} \cdot n_{C_2H_4} + Q_{p\,C_2H_6} \cdot n_{C_2H_6}}{1000} \tag{2}$$

W—calorific value (MJ/m^3);
Q_p—heat of combustion (kJ/m^3); and
n—mole fractions CO, CH$_4$, C$_2$H$_2$, C$_2$H$_4$, and C$_2$H$_6$.

For the calculation of specific energy input (*SEI*) and energy efficiency (*EE*), the following equations were used [34]:

$$SEI = \frac{P}{Q} \tag{3}$$

SEI—specific energy input (kwh/m^3),
P—discharge power (kW), and
Q—total flow rate (m^3/h).

$$EE = \frac{Co - C}{SEI} \tag{4}$$

EE—energy efficiency (g/kWh) and
SEI—specific energy input (kWh/m^3).

4. Reaction Mechanism

The catalysts were rinsed with acetone after the process; then, the solution obtained was analyzed with MS to identify toluene decomposition products (Figure 5). Retention times (RTs) between 0.48 and 0.54 min were methanol, acetone, water, and C2-C4 hydrocarbons. 2-Pentanone was identified (RT = 2.75 min) as a product of the acetone condensation reaction. The proposed reaction mechanism was based on previous [20,21,24] and current studies' toluene decomposition intermediate products as well as decomposition mechanisms proposed in the literature [27,31–36]. The process was inducted by excited electrons, which could cause both radical and electron dissociation of the particles. Electron attachment to toluene particles could lead to the formation of CH$_3$ or H radials. Since the gas consisted of nitrogen, carbon oxides, and hydrogen, those could also be dissociated by the impact of electrons [21]. Water was formed during the methanation reaction. The presence of humidity in nitrogen-rich gas was a source of HO• and NO• radicals, but a higher density of hydrogen oxide radicals was observed [37]. The reason could be that the source of nitrogen radicals, the N≡N bond, required more energy to break (9.8 eV) than the H-OH bond in the water molecule (5.11 eV) [35].

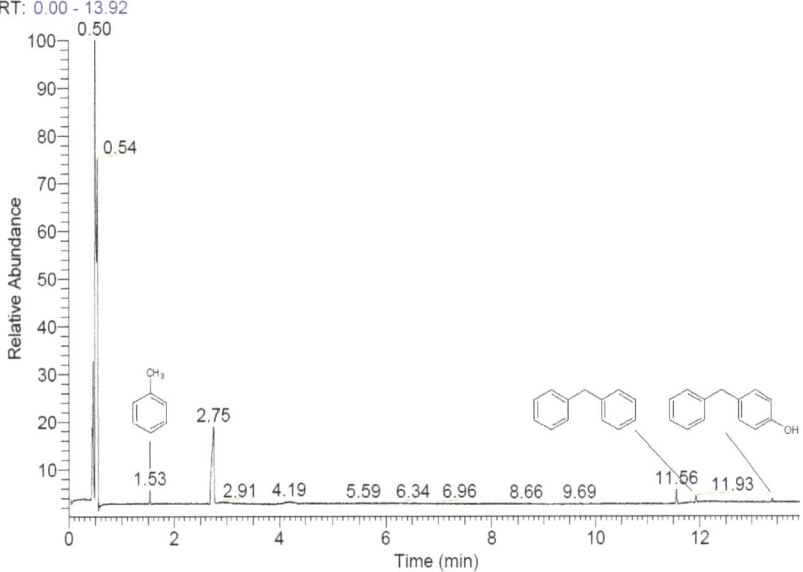

Figure 5. MS analysis of Ni/CaO-Al$_2$O$_3$ catalyst surface rinsed with acetone after the process.

No toluene decomposition reaction products were present on the surface of NiO/(CaO-Al$_2$O$_3$) catalysts, and only a trace amount of C$_7$H$_8$ (RT = 1.53 min) was adsorbed (Table 3). No nitrogen oxides in gas after reaction or nitrogen-containing intermediate products such as nitrotoluene on catalysts' surfaces were identified for all of the systems studied.

Table 3. Compounds adsorbed on catalysts surface after the process [24].

Substance	NiO/(CaO-Al$_2$O$_3$)	Ni/(CaO-Al$_2$O$_3$)	NiO/Al$_2$O$_3$ [24]
Toluene	+	+	+
Methanol	−	+	+
3-hexen-2-one	−	−	+
Diphenylmethane	−	+	−
4-Hydroxybenzophenone	−	+	−

On the nickel oxide catalyst without calcium addition, only 3-hexen-2-one was identified as a product of the toluene ring-opening reaction (5) (Figure 6). On the metallic nickel catalyst, tar formation was observed despite CaO addition in the catalyst carrier. A higher hydrogen concentration in the initial gas and an increase in discharge power resulted in a higher amount of hydrogen radicals, which could then react with other particles during the process. In the current study, the deactivation of Ni/CaO-Al$_2$O$_3$ could be caused by too high of a Ni/Ca ratio on the catalyst bed, which led to a decrease in the number of active sites on the catalyst. A higher amount of hydrogen radicals did not lead to a higher conversion rate or resistance to deactivation in previous studies [24]. Instead, the radicals were used in the hydrogenation of toluene decomposition intermediates due to a lack of the active sites on the catalyst surface, which lowered the C$_7$H$_8$ conversion rate.

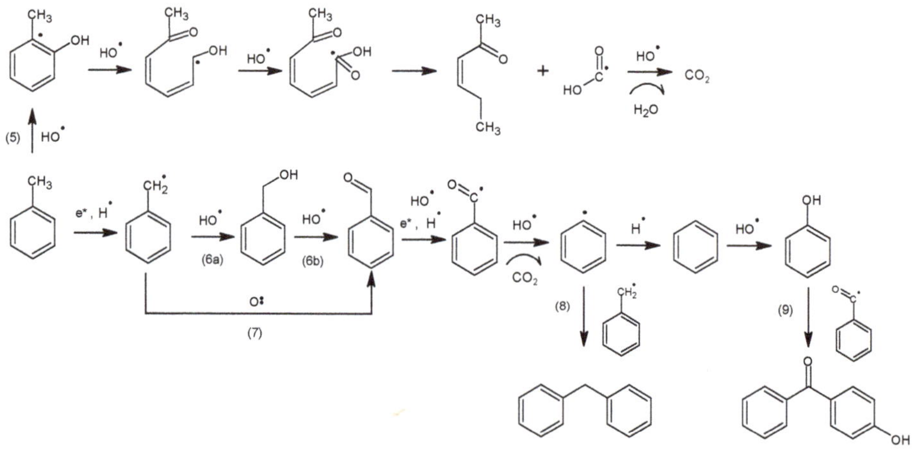

Figure 6. Possible reactions of toluene in plasma–catalytic system.

In this study, polycyclic hydrocarbons such as diphenylmethane (RT = 11.93 min) were identified on the catalyst surface. The formation of benzyl alcohol identified in a previous study [24] was a product of oxidation reactions (6a and 6b) with HO• or oxide (7) radicals, which could also lead to the formation of other intermediates such as benzoic acid, benzene [20,21], or phenol [34] as a result of further oxidation reactions [35]. Intermediate products of toluene decomposition further reacted with each other [36] to form polycyclic hydrocarbons (8 and 9), which were identified during the studies. Radical and electrons reacted with toluene decomposition products to form simple particles such as hydrogen, water, carbon oxide, and dioxide [24]. A trace amount of methanol identified at RT 0.48 min (Figure 5) on the catalyst carrier was formed during methane oxidation [38].

For each catalyst, a similar gas composition was observed after the process. In the outlet gas, trace amounts of C2 hydrocarbons were formed. The concentration of hydrogen was lowered mainly due to the methanation process and hydrogenation of toluene decomposition products. In the presence of plasma or excited molecules, CO_2 could decompose to CO and oxygen or oxide radicals [21], which then reacted with hydrogen to form water. CO_2 could not undergo the methanation process because its reaction is strongly inhibited by CO methanation. Moreover, the reaction of hydrogen with CO was twice as fast as with CO_2 [39]. During the process, the concentration of CO increased, although it was used for methane production. This also led to the conclusion that it was a product of the decomposition of carbon dioxide. This reaction had a favorable effect on the calorific value of the outlet gas.

Plasma–catalyst modeling, focused on the effects of both the catalysts surface and plasma effects, has not been extensively studied due to the complexity of the subject matter. However, several limited models have been reported. The mechanism of the chemical reactions involved in the removal of NF_3 in a hybrid plasma $BaTiO_3$ packed-bed with a $CaCO_3$ absorbent was modeled. Theoretically predicted byproducts were successfully identified by the authors during the experiments. The removal efficiency of NF_3 was improved with a $BaTiO_3$ dielectric constant enhancing electric field but decreased with the increase in gas flow [40]. A negative influence of gas flow on the conversion rate was also found when comparing the results presented in this paper with other studies [31–33]. Another group used fluid modeling and Particle-in-cell/Monte Carlo collision (PIC/MCC) to study the electric field enhancement and micro discharge formation in catalysts' pores on the catalyst–DBD system. Plasma formation in the pores occurred more easily in common catalysts supports such as Al_2O_3 with low dielectric constant than in ferroelectric catalysts. This resulted in a larger contact area for plasma and catalysts and could improve the plasma–catalyst process [41]. On the contrary, the effect of electric field enhancement was more prominent in materials with higher dielectric constant, which could lead to stronger oxidative power of plasma discharge [42]. It became clear that further studies on the plasma–catalytic mechanisms are needed as this area has not yet been studied in depth.

5. Conclusions

In plasma–catalytic systems with nickel catalysts deposited on Al_2O_3 and CaO-Al_2O_3, an efficient tar decomposition process was possible. High conversion of C_7H_8 was observed—up to 85%, which exceeded the results obtained without the catalyst.

An increase in hydrogen concentration resulted in a higher conversion of toluene when the Ni/(CaO-Al_2O_3) catalyst was used. Hydrogen was consumed in the methanation reaction and water formation. These reactions did not significantly affect the calorific value of the gas. With the addition of CaO, the catalyst bed decreased the amount of toluene decomposition products adsorbed on NiO/(CaO-Al_2O_3) catalyst surface.

Author Contributions: Conceptualization, J.W.-W. and M.M.; Formal analysis, B.U.; Funding acquisition, K.K.; Investigation, J.W.-W. and M.M.; Methodology, J.W.-W.; Visualization, J.W.-W.; Writing—original draft, J.W.-W.; Writing—review and editing, M.M. All authors have read and agreed to the published version of the manuscript.

Funding: This research was funded by the National Center for Research and Development agreement no. PBS2/A1/10/2013.

Conflicts of Interest: The authors declare no conflict of interest.

References

1. Zhang, Z.; Pang, S. Experimental investigation of tar formation and producer gas composition in biomass steam gasification in a 100kW dual fluidized bed gasifier. *Renew. Energy* **2019**, *132*, 416–424. [CrossRef]
2. Di Carlo, A.; Borello, D.; Sisinni, M.; Savuto, E.; Venturini, P.; Bocci, E.; Kuramoto, K. Reforming of tar contained in a raw fuel gas from biomass gasification using nickel-mayenite catalyst. *Int. J. Hydrogen Energy* **2015**, *40*, 9088–9095. [CrossRef]
3. Heidenreich, S.; Foscolo, P. New concepts in biomass gasification. *Prog. Energy Combust. Sci.* **2015**, *46*, 72–95. [CrossRef]
4. Qin, Y.; Campen, A.; Wiltowski, T.; Feng, J.; Li, W. The influence of different chemical compositions in biomass on gasification tar formation. *Biomass Bioenergy* **2015**, *83*, 77–84. [CrossRef]
5. Bhaduri, S.; Contino, F.; Jeanmart, H.; Breuer, E. The effects of biomass syngas composition, moisture, tar loading and operating conditions on the combustion of a tar-tolerant HCCI (Homogeneous Charge Compression Ignition) engine. *Energy* **2015**, *87*, 289–302. [CrossRef]
6. Das, S.; Kashyap, D.; Kalita, P.; Kulkarni, V.; Itaya, Y. Clean gaseous fuel application in diesel engine: A sustainable option for rural electrification in India. *Renew. Sustain. Energy Rev.* **2020**, *117*, 109485. [CrossRef]
7. Fjellerup, J.; Ahrenfeldt, J.; Henriksen, U.; Gobel, B. *Formation, Decomposition and Cracking of Biomass Tars in Gasification*; Technical University of Denmark, Department of Mechanical Engineering: Kgs. Lyngby, Denmark, 2005; pp. 1–60.
8. Shen, Y. Chars as carbonaceous adsorbents/catalysts for tar elimination during biomass pyrolysis or gasification. *Renew. Sustain. Energy Rev.* **2015**, *43*, 281–295. [CrossRef]
9. Richardson, Y.; Blin, J.; Julbe, A. A short overview on purification and conditioning of syngas produced by biomass gasification: Catalytic strategies, process intensification and new concepts. *Prog. Energy Combust. Sci.* **2012**, *38*, 761–781. [CrossRef]
10. Riosa, M.-L.-V.; González, A.-M.; Lora, E.-E.-S. Almazán del Olmoc OA. Reduction of tar generated during biomass gasification: A review. *Biomass Bioenergy* **2018**, *108*, 345–370.
11. Huang, Z.; Zheng, A.; Deng, Z.; Wei, G.; Zhao, K.; Chen, D.; He, F.; Zhao, Z.; Lia, H.; Li, F. In-situ removal of toluene as a biomass tar model compound using $NiFe_2O_4$ for application in chemical looping gasification oxygen carrier. *Energy* **2020**, *190*, 116360. [CrossRef]
12. Paethanom, A.; Nakahara, S.; Kobayashi, M.; Prawisudha, P.; Yoshikawa, K. Performance of tar removal by absorption and adsorption for biomass gasification. *Fuel Process. Technol.* **2012**, *104*, 144–154. [CrossRef]
13. Zhang, Z.; Liu, L.; Shen, B.; Wu, C. Preparation, modification and development of Ni-based catalysts for catalytic reforming of tar produced from biomass gasification. *Renew. Sust. Energy Rev.* **2018**, *94*, 1086–1109. [CrossRef]
14. Xie, Y.; Su, Y.; Wang, P.; Zhang, S.; Xiong, Y. In-situ catalytic conversion of tar from biomass gasification over carbon nanofibers-supported Fe-Ni bimetallic catalysts. *Fuel Process. Technol.* **2018**, *182*, 77–87. [CrossRef]
15. Duman, G.; Watanabe, T.; Uddin, M.-A.; Yanik, J. Steam gasification of safflower seed cake and catalytic tar decomposition over ceria modified iron oxide catalysts. *Fuel Process. Technol.* **2014**, *126*, 276–283. [CrossRef]
16. Tan, R.-S.; Abdullah, T.-A.-T.; Ripin, A.; Ahmad, A.; Isac, K.M. Hydrogen-rich gas production by steam reforming of gasified biomass tar over Ni/dolomite/La_2O_3 catalyst. *J. Environ. Chem. Eng.* **2019**, *7*, 103490. [CrossRef]

17. Wang, Y.; Yang, H.; Tu, X. Plasma reforming of naphthalene as a tar model compound of biomass gasification. *Energy Convers. Manag.* **2019**, *187*, 593–604. [CrossRef]
18. Liu, L.; Zhang, Z.; Das, S.; Kawi, S. Reforming of tar from biomass gasification in a hybrid catalysis-plasma system: A review. *Appl. Catal. B* **2019**, *250*, 250–272. [CrossRef]
19. Saleem, F.; Zhang, K.; Harvey, A. Plasma-assisted decomposition of a biomass gasification tar analogue into lower hydrocarbons in a synthetic product gas using a dielectric barrier discharge reactor. *Fuel* **2019**, *235*, 1412–1419. [CrossRef]
20. Woroszył-Wojno, J.; Młotek, M.; Perron, M.; Jóźwik, P.; Ulejczyk, B.; Krawczyk, K. Decomposition of Tars on a Nickel Honeycomb Catalyst. *Catalysts* **2021**, *11*, 860. [CrossRef]
21. Młotek, M.; Woroszył, J.; Ulejczyk, B.; Krawczyk, K. Coupled Plasma-Catalytic System with Rang 19PR Catalyst for Conversion of Tar. *Sci. Rep.* **2019**, *9*, 13562. [CrossRef]
22. Młotek, M.; Ulejczyk, B.; Woroszył, J.; Walerczak, I.; Krawczyk, K. Purification of the gas after pyrolysis in coupled plasma-catalytic system. *Pol. J. Chem. Technol.* **2017**, *19*, 94–98. [CrossRef]
23. Młotek, M.; Ulejczyk, B.; Woroszył, J.; Krawczyk, K. Decomposition of Toluene in Coupled Plasma-Catalytic System. *Ind. Eng. Chem. Res.* **2020**, *59*, 4239–4244. [CrossRef]
24. Woroszył-Wojno, J.; Młotek, M.; Ulejczyk, B.; Krawczyk, K. Nickel catalyst in coupled plasma-catalytic system for tar removal. *Pol. J. Chem. Technol.* **2021**, *23*, 24–29. [CrossRef]
25. Tao, K.; Ohta, N.; Liu, G.; Yoneyama, Y.; Wang, T.; Tsubaki, N. Plasma enhanced catalytic reforming of biomass tar model compound to syngas. *Fuel* **2013**, *104*, 53–57. [CrossRef]
26. Hou, Z.; Yokota, O.; Tanaka, T.; Yashima, T. Characterization of Ca-promoted Ni/α-Al$_2$O$_3$ catalyst for CH$_4$ reforming with CO$_2$. *Appl. Catal. A* **2003**, *253*, 381–387. [CrossRef]
27. Li, C.; Hirabayashi, D.; Suzuki, K. Development of new nickel based catalyst for biomass tar steam reforming producing H$_2$-rich syngas. *Fuel Process. Technol.* **2009**, *90*, 790–796. [CrossRef]
28. Franczyk, E.; Gołębiowski, A.; Borowiecki, T.; Kowalik, P.; Wróbel, W. Influence of Steam Reforming Catalyst Geometry On The Performance Of Tubular Reformer—Simulation Calculations. *Chem. Eng. Process.* **2015**, *36*, 239–250. [CrossRef]
29. Ashok, J.; Kathiraser, Y.; Ang, M.-L.; Kawi, S. Bi-functional hydrotalcite-derived NiO–CaO–Al$_2$O$_3$ catalysts for steam reforming of biomass and/or tar model compound at low steam-to-carbon conditions. *Appl. Catal. B* **2015**, *172–173*, 116–128. [CrossRef]
30. Choong, C.-K.-S.; Zhong, Z.; Huang, L.; Wang, Z.; Ang, T.-P.; Borgna, A.; Lin, J.; Hong, L.; Chen, L. Effect of calcium addition on catalytic ethanol steam reforming of Ni/Al$_2$O$_3$: I. Catalytic stability, electronic properties and coking mechanism. *Appl. Catal. A* **2011**, *407*, 145–154. [CrossRef]
31. Meia, D.; Liu, S.; Wang, Y.; Yang, H.; Bo, Z.; Tu, X. Enhanced reforming of mixed biomass tar model compounds using a hybrid gliding arc plasma catalytic process. *Catal. Today* **2019**, *337*, 225–233. [CrossRef]
32. Kong, X.; Zhang, H.; Li, X.; Xu, R.; Mubeen, I.; Li, L.; Yan, J. Destruction of Toluene, Naphthalene and Phenanthrene as Model Tar Compounds in a Modified Rotating Gliding Arc Discharge Reactor. *Catalysts* **2019**, *9*, 19. [CrossRef]
33. Yu, L.; Li, X.; Tu, X.; Wang, Y.; Lu, S.; Yan, J. Decomposition of Naphthalene by dc Gliding Arc Gas Discharge. *J. Phys. Chem. A* **2010**, *114*, 360–368. [CrossRef]
34. Liu, S.; Mei, D.; Tu, X. Steam reforming of toluene as biomass tar model compound in a gliding arc discharge reactor. *Chem. Eng. J.* **2017**, *307*, 793–802. [CrossRef]
35. Du, C.-M.; Yan, J.-H.; Cheron, B. Decomposition of toluene in a gliding arc discharge plasma reactor. *Plasma Sources Sci. Technol.* **2007**, *16*, 791–797. [CrossRef]
36. Colket, M.-B.; Seery, D.-J. Reaction mechanisms for toluene pyrolysis. *Symp. Int. Combust. Proc.* **1994**, *25*, 883–891. [CrossRef]
37. Benstaali, B.; Boubert, P.; Cheron, B.-G.; Addou, A.; Brisset, J.-L. Density and Rotational Temperature Measurements of the OH° and NO° Radicals Produced by a Gliding Arc in Humid Air. *Plasma Chem. Plasma Process.* **2002**, *22*, 553–571. [CrossRef]
38. Aghamir, F.-M.; Matin, N.-S.; Jalili, A.-H.; Esfarayeni, M.-H.; Khodagholi, M.-A.; Ahmadi, R. Conversion of methane to methanol in an ac dielectric barrier discharge. *Plasma Sources Sci. Technol.* **2004**, *13*, 707–711. [CrossRef]
39. Gołębiowski, A.; Stołecki, K. The kinetics of methanation of CO and CO$_2$ under industrial conditions. *Przem. Chem.* **2001**, *80*, 514–516.
40. Chang, J.-S.; Kostov, K.-G.; Urashima, K.; Yamamoto, T.; Okayasu, Y.; Kato, T.; Iwaizumi, T.; Yoshimura, K. Removal of NF$_3$ from semiconductor-process flue gases by tandem packed-bed plasma and adsorbent hybrid systems. *IEEE Trans. Ind. Appl.* **2000**, *36*, 1251–1259. [CrossRef]
41. Bogaerts, A.; Zhang, Q.-Z.; Zhang, Y.-R.; Laer, K.-V.; Wang, W. Burning questions of plasma catalysis: Answers by modeling. *Catal. Today* **2019**, *337*, 3–14. [CrossRef]
42. Holzer, F.; Kopinke, F.-D.; Roland, U. Influence of Ferroelectric Materials and Catalysts on the Performance of Non-Thermal Plasma (NTP) for the Removal of Air Pollutants. *Plasma Chem. Plasma Process.* **2005**, *25*, 595–611. [CrossRef]

MDPI
St. Alban-Anlage 66
4052 Basel
Switzerland
Tel. +41 61 683 77 34
Fax +41 61 302 89 18
www.mdpi.com

Catalysts Editorial Office
E-mail: catalysts@mdpi.com
www.mdpi.com/journal/catalysts

www.ingramcontent.com/pod-product-compliance
Lightning Source LLC
LaVergne TN
LVHW070627100526
838202LV00012B/746